FIRST BITE

By the same author:

Consider the Fork:
A History of How We Cook and Eat

The Hive:
The Story of the Honeybee and Us

Swindled: The Dark History of Food Fraud,
from Poisoned Candy to Counterfeit Coffee

FIRST BITE

HOW WE LEARN TO EAT

BEE WILSON

Illustrations by ANNABEL LEE

FOURTH ESTATE · London

First published in Great Britain by
Fourth Estate
an imprint of HarperCollinsPublishers
1 London Bridge Street
London SE1 9GF
www.4thestate.co.uk

A catalogue record for this book is available from the British Library

ISBN 978-0-00-754970-2

Printed in Italy by Grafica Veneta S.p.A.

MIX
Paper from
responsible sources
FSC™ C007454

For Emily

Contents

Preface		ix
Introduction		1
CHAPTER 1:	Likes and Dislikes	27
	BEETROOT	70
CHAPTER 2:	Memory	73
	MILK	107
CHAPTER 3:	Children's Food	109
	BIRTHDAY CAKE	148
CHAPTER 4:	Feeding	151
	LUNCHBOX	187
CHAPTER 5:	Brothers and Sisters	189
	CHOCOLATE	223
CHAPTER 6:	Hunger	225
	BREAKFAST CEREAL	260
CHAPTER 7:	Disorder	263
	CRISPS	303
CHAPTER 8:	Change	305
	CHILLI	348
EPILOGUE:	This is Not Advice	351
	Further Reading	357
	Bibliography	359
	Notes	391
	Acknowledgements	401
	Index	405

Preface

Some find the whole matter of eating easy, while others find it hard. I used to be on the wrong side of this great divide and somehow, to my own surprise and relief, leaped over to the other side. This book is my attempt to explore how this switch was possible.

You don't have to look far in our world to encounter people – of all sizes – who relate to food in chaotic ways. The chaos can take many forms: compulsive overeating, undereating, or extreme pickiness. Some people become so obsessed with the purity of what enters their mouths that they cannot accept invitations to eat with friends. It is a lonely occupation, being someone who wrestles to control their responses to food, given that modern life is steeped with things to eat, both real and imaginary. Snacks assail us at the checkout; dream feasts tease us from hoardings, newspapers and TV cooking shows.

Without ever quite having a full-blown eating disorder – though I came close – I managed to make myself pretty miserable about eating for the best part of a decade, from the middle school years to young adulthood. I probably appeared fine: a bit overweight, nothing more. But food was my main relationship, and – although it had some of the thrills of romance, especially when I was in the kitchen with a hunk of sweet brioche dough – it wasn't a stable or sustaining kind of love. We talk in a sickly way of 'indulgent' foods, but when you are trapped in compulsive habits of dependency on them, it does not feel like being pampered. There were days when I gave myself up to consuming guilty treats. Other days were for not eating, which was even worse, taunting myself with the foods that I wouldn't permit myself.

Thankfully, that phase of life now seems distant. Eating well – by which I don't mean 'clean eating' or raw juice fasts but regular meals of real, flavoursome food – just isn't that complicated for me any more. Now that I am through on the other side, I can see that over a period of months, if not years, I learned to master a series of skills that I'd once deemed insurmountable. I learned that it was OK to eat a hearty meal when I was hungry; but also OK to stop when I was full. My cravings for pastries lessened and my cravings for vegetables increased. There are still plenty of things I worry and obsess about, believe me, but my own eating is seldom one of them. Dinner is just dinner: nothing more nor less than the high point of the day.

In our house, as in many others, the battleground over food has shifted to the children. As a parent trying to get my three children to eat healthily – but not obsessively so – I have sometimes felt as lost as I once did about my own eating. After the milk stage (and that was hard enough) none of the skills of feeding came naturally. How do you promote vegetables to an ironic teenager in a way that isn't counter-productive? What do you do when your daughter comes home and says her friends have started skipping lunch? How do you keep a sense of proportion about fat and sugar without giving in completely to the ultra-processed food that is now ubiquitous?

In those busy moments after school and before bed, I cook a quick meal that I hope will please everyone. I may find that one child grumbles about the grilled aubergines, another says they are the best bit and the third sits quietly weeping because, while he actually likes aubergines, they are touching a piece of chicken and are therefore inedible. Did I say the high point of the day? And yet, comparatively speaking, my children are not problem eaters.

All parents have moments of thinking that it just isn't possible to teach a child to eat well; or at least not *your* child. Many grown-ups are more pessimistic still about their own ability to change how they behave around food. But writing this book has taught me that there is immense potential for improving our eating habits. It may take longer for some people to get there than others, but learning how to eat better – which is quite different from going on a diet – is within anyone's grasp. Perhaps the most eloquent argument for learning new ways to eat is that of pleasure. Eating is – or should be – a daily source of delight rather than something to fight against. It's good here, on the other side of the divide. I do hope you'll join me.

Introduction

'One of the reasons I like bread and jam,' said
Frances, 'is that it does not slide off your spoon
in a funny way.'

RUSSELL HOBAN, *Bread and Jam for Frances*

So many of our anxieties around diet take the form of a
search for the perfect food, the one that will cure all our ills.
Eat this! Don't eat that! We obsess about the properties of
various ingredients: the protein, the omega oils, the vitamins.
But this is getting ahead of ourselves. Nutrients only count
when a person picks up food and eats it. *How* we eat – how
we approach food – is what really matters. If we are going
to change our diets, we first have to relearn the art of eating,
which is a question of psychology as much as nutrition. We
have to find a way to want to eat what's good for us.

Our tastes follow us around like a comforting shadow.
They seem to tell us who we are. Maybe this is why we act
as if our core attitudes to eating are set in stone. We make
frequent attempts – more or less half-hearted – to change
what we eat, but almost no effort to change how we feel
about food: how well we deal with hunger, how strongly

attached we are to sugar, our emotions on being served a small portion. We try to eat more vegetables, but we do not try to make ourselves enjoy vegetables more, maybe because there's a near-universal conviction that it is not possible to learn new tastes and shed old ones. Yet nothing could be further from the truth.

All the foods that you regularly eat are ones that you learned to eat. Everyone starts life drinking milk. After that, it's all up for grabs.

Bone marrow from wild game is considered the best first baby food among the hunting tribes of Tanzania.[1] If you were born in the Far Eastern republic of Laos, it could be gelatinous rice, pre-chewed by your mother and transferred from her mouth to yours (this is sometimes called kiss-feeding).[2] For Western babies, that first bite of solid food may be powdered cereal from a packet or purée from a jar; it could be organic pumpkin, steamed and strained and served with a hypoallergenic spoon; or a random nibble from a parent's plate. Aside from milk, there is simply no such thing as a universal food. Not even for babies.*

---

* Even milk is complicated. Formula will never be the same as breast milk, as the breastfeeding campaigners often remind us. But nor is human milk a single substance. It's been found that breastfed babies in Spain have a different range of bacteria in their gut from breastfed babies in Sweden. A mother's milk will vary in composition and flavour depending on her own diet. It may taste garlicky in France or be scented with star anise in China. Slightly surprisingly, not everyone recognizes it as the ideal food for newborns. Let's go back to my sentence about us all starting life drinking milk. It isn't quite true. There are remote rural cultures where they believe that babies will be harmed by colostrum, the rich yellowish milk that mothers produce in the first few days after birth. Parents may give babies honey or sweet almond oil for the first three days instead, because they fear – wrongly – that this early milk is too 'strong' for a tiny baby to digest.

From our first year of life, human tastes are astonishingly diverse. As omnivores, we have no inbuilt knowledge of which foods are good and safe. Each of us has to use our senses to figure out for ourselves what is edible, depending on what's available. In many ways, this is a delightful opportunity. It's the reason there are such fabulously varied ways of cooking in the world.

But we haven't paid anything like enough attention to another consequence of being omnivores, which is that eating is not something we are born instinctively knowing how to do, like breathing. It is something we learn. A parent feeding a baby is training them how food should taste. At the most basic level, we have to learn what is food and what is poison. We have to learn how to satisfy our hunger and also when to stop eating. Unlike the anteater which eats only small termites, we have few natural instincts to fall back on. Out of all the choices available to us as omnivores, we have to figure out which foods are likeable, which are lovable and which are disgusting. From these preferences, we create our own pattern of eating, as distinctive as a signature.

Or that's how it used to be. In today's food culture, many people seem to have acquired uncannily homogenous tastes, markedly more so than in the past. In 2010 two consumer scientists argued that the taste preferences of childhood provided a new way of thinking about the causes of obesity. They noted a 'self-perpetuating cycle': food companies push foods high in sugar, fat and salt, which means that children learn to like them, and so the companies invent ever more of these foods 'that contribute to unhealthy eating habits'.[3] The main influence on a child's palate may no longer be a parent but a series of food manufacturers whose products – despite their illusion of infinite choice – deliver a monotonous

flavour hit, quite unlike the more varied flavours of traditional cuisine.

I went to the cinema with one of my children recently. We
stood at the ice-cream concession and I realized, with a jolt,
that almost all the options – other than plain vanilla – contained chocolate in one form or another. Would we pick mint
chocolate chunk or cherry chocolate chunk or chocolate ice
cream with chocolate brownie pieces or caramel ice cream
with pieces of caramel chocolate? The danger of growing up
surrounded by these endless sweet and salty industrial concoctions is not that we are innately incapable of resisting
them but that the more frequently we eat them, especially in
childhood, the more they train us to expect all food to taste
this way.

Once you recognize the simple fact that food preferences
are learned, many of the ways we currently approach eating
start to look a little weird. To take a small example, consider the parents who go to great lengths to 'hide' vegetables
in children's meals. Is broccoli really so terrible that it must
be concealed from innocent minds? Whole cookbooks have
been devoted to this arcane pursuit. It starts with the notion
that children have an innate resistance to vegetables, and
will only swallow them unawares, blitzed into pasta sauce or
baked into sweet treats; they could never learn to love courgette for its own sake. In our harried, sleep-deprived state,
parents find it hard to play the long game. We think we are
being clever when we smuggle some beetroot into a cake. Ha!
Tricked you into eating root vegetables! But since the child is
not conscious that they are consuming beetroot, the main up-
shot is to entrench their liking for cake. A far cleverer thing
would be to help children learn to become adults who choose
vegetables consciously, of their own accord.

By failing to see that eating habits are learned, we misunderstand the nature of our current diet predicament. As we are often reminded, in doom-laden terms, eating has taken a dramatic collective wrong turn in recent decades. As of 2010, poor diet and physical inactivity accounted for 10 per cent of all deaths and disease worldwide, ahead of tobacco smoke (6.3 per cent) and household air pollution (4.3 per cent).[4] Around two-thirds of the population are either overweight or obese in rich countries; and the rest of the world is fast catching up. The moral usually drawn from these statistics is that we are powerless to resist the sugary, salty, fatty foods that the food industry promotes. Everything tastes better with bacon! As the journalist Michael Moss exposed in 2013, the big food companies engineer foods with a chemically calculated 'bliss point' designed to get us hooked.[5] Newspapers sometimes project a future in which obesity levels continue to rise indefinitely until almost everyone in the world is affected.

But there's something else going on here, which usually gets missed. Not everyone is equally susceptible to the dysfunction of our food supply. Some people manage to eat sugary, salty, fatty foods in modest quantities, and then stop. Others find these supposedly irresistible foods the opposite of blissful. If two-thirds of the population are overweight or obese, then fully a third are not. This is astonishing, given just how many opportunities there now are to eat doughnuts. Exposed to the same food that bombards us all, these lucky people have learned different responses. It's in all our interests to find out how they have done it.

Many campaigners would say cooking is the answer. If only children could be taught how to cook and plant vegetable gardens, they would automatically become healthier. It sounds convincing: school gardens are a lovely thing. But by themselves, they are not enough to make a child relate to food in healthy ways. Our difficulty is not just that we haven't learned to cook and grow food, however important that is: it's that we haven't learned to *eat* in ways that support health and happiness. Traditional cuisines across the world were founded on a strong sense of balance, with norms about which foods go together, and how much one should eat at different times of day. Much cooking now, however, is nothing like this. In my experience as a food journalist, chefs and food writers are, if anything, more prone to compulsive eating and other disordered food obsessions than non-cooks. For cooking to become the solution to our diet crisis, we first have to learn how to adjust our responses to food. Cooking skills are no guarantee of health if your inclinations are for twice-fried chicken, Neapolitan rum babas and French aligot: potatoes mashed with a ton of cheese.

The reason many find it hard to eat healthily is that we have never learned any differently. Like children, most of us eat what we like and we only like what we know. Never before have whole populations learned (or mislearned) to eat in societies where calorie-dense food was so abundant and policed with so few norms about portion sizes and mealtimes. Nor is overeating the only problem that plagues modern affluent civilizations. Statistics suggest that around 0.3 per cent of young women are anorexic and another 1 per cent are bulimic, with rising numbers of men joining them.[6] What statistics are not particularly effective at telling us is how many others – whether overweight or underweight – are in a perpetual state of anxiety about what they consume, living in fear of carbs or fat grams and unable to derive straightforward enjoyment from meals. A 2003 study of 2,200

American college students suggested that weight concern is very common: 43 per cent of this sample were worried about their weight most of the time (across both sexes) and 29 per cent of the women described themselves as 'obsessively preoccupied' with weight.[7]

Our dietary malaise is often discussed in fatalistic terms, as if our preference for hamburgers were a life sentence: diets don't work, sugar is addictive and so on. What we forget is that, as omnivores, we are extremely gifted at changing the way we eat to accommodate different environments. Admittedly, no one has ever encountered a food environment quite like the one we now find ourselves in, flooded with cheap calories in deceptive packaging. To survive in our current situation will entail very different skills from those needed by a Palaeolithic hunter-gatherer. Yet there is every reason to suppose that we are capable of acquiring these skills, if we give ourselves half a chance.

If our food habits are learned, they can also be relearned. Imagine you were adopted at birth by parents who lived in a remote village in a far-flung country. Your tastes would be quite unlike the ones you ended up with. We all begin life with an innate liking for sweetness and a suspicion of bitterness, yet there is nothing inevitable in our physiology that says that we will grow up dreading vegetables and craving fudge. The trouble is, we do not tend to see it this way.

My premise in *First Bite* is that the question of how we learn to eat – both individually and collectively – is the key to how food, for so many people, has gone so badly wrong. The greatest public health problem of modern times is how to persuade people to make better food choices. But we have been looking for answers in the wrong places.

Our discussion about diet is usually framed in terms of better information. A sea of articles and books suggests that the reason for the obesity crisis is that we were given the wrong advice: we were told to avoid fat when the real demon was sugar.[8] There's something in this. It certainly didn't help that many of the 'low-fat' products marketed as healthy over the past few decades were padded with refined carbohydrates and were therefore more fattening than the fats we were being advised to give up.[9] During the period that dieticians were admonishing us to abstain from saturated fats such as those in butter, cream and meat, obesity rates were consistently going up, not down. It is becoming increasingly clear that eating fat is not in and of itself the thing that makes you fat or gives you heart disease.

Before we start blaming the confusing low-fat advice for our current ill health, however, it might be useful to consider the extent to which we ever followed those anti-fat warnings. The vast majority of people heard what the 'food police' had to say on the subject of fat and ignored it. At the height of the low-fat orthodoxy, in 1998, some of the leading nutrition scientists in the world co-authored a paper in which they lamented the public's failure to follow their guidelines. The scientists found, to their dismay, that after more than two decades of being advised to reduce fat, people were still eating 'about the same' amount of it. The percentage of calories from fat in the American diet dropped slightly from 1976 to 1991 (from about 36 per cent in 1976 to 34 per cent in 1991) but this was only because people were consuming more total calories. In absolute terms, the fat grams people consumed on average remained the same.[10]

David L. Katz of the Yale University Prevention Research Center is a rare voice of sanity in the clamorous world of nutrition. He disputes the commonly held view that the reason we don't eat better is because there is so much confusion over what the 'best diet' really is. Katz points out that the essential tenets of a healthy life – moderate helpings of a variety of real whole foods, plus regular exercise – have been very well established for decades. The medical evidence suggests that it doesn't matter whether we reach this point via a low-fat route or a low-carb one (or vegan or Paleo or just good old-fashioned home cooking).[11] Across all diets, there is, notes Katz, a huge 'aggregation of evidence' that the best pattern of eating for health is a diet of minimally processed foods, mostly plant-based. 'Our problem,' notes Katz, 'is not want of knowledge about the basic care and feeding of Homo Sapiens. Our problem is a stunning and tragically costly cultural reluctance – to swallow it.'[12]

Take vegetables. The advice to eat more vegetables for health could hardly have been clearer. We have been given the message many times, in many forms. Unlike with fat or sugar, there is no about-turn or controversy in mainstream nutrition over the 'eat more vegetables' message. Yet since the 1970s the total intake of calories from vegetables in America has actually declined by 3 per cent, which is a bigger drop than it sounds considering that vegetables contain very few calories compared to other foods.[13] This decline came at a time when there has never been a wider variety of tempting vegetables available, from deep orange butternut squash to pale green Romanesco broccoli. Many people, however, have absorbed the lesson from childhood that vegetables and pleasure – and more generally, healthy food and pleasure – can never go together. Witness the waves of antipathy directed at public figures such as Michelle Obama when they dare to suggest that we eat more of them. Consumer scientists have found that when a new product is described as 'healthy', it is far less likely to be a success than if it is described as 'new'.[14]

When it comes to our dining habits, there is a giant mismatch between thought and deed; between knowledge and behaviour. 'Eat food. Not too much. Mostly plants,' says influential food writer Michael Pollan.[15] A wise and simple mantra, much repeated; yet for many it seems anything but simple to follow in daily life. To adhere to it you need to: 'Like real food. Not enjoy feeling overstuffed. And appreciate vegetables.' These are skills that many people have not yet acquired, however intelligent or advanced in years they may be. There's another complication, too. The 'not too much' part of Pollan's dictum needs modifying to take account of those who have learned to eat *too little*, or at least not enough of the right foods. I am not just talking about the underweight. The term 'malnutrition' now covers obesity as well as starvation; there is evidence that obese populations across the world suffer disproportionately from micronutrient deficiencies, notably vitamins A and D, plus zinc and iron.[16] Learning how to eat better is not about reducing consumption across the board. While we undoubtedly need to eat less of many foods – sugar springs to mind – we need more of others. Among other lost eating skills – see also not 'spoiling your appetite' and not 'wolfing down your dinner' – we seem to have lost the old-fashioned concept of 'nourishing' ourselves.

A tone of judgemental impatience often creeps into discussions of obesity. 'It's not exactly rocket science, is it?' is a frequent observation on newspaper comment boards, from some of those lucky ones who have never struggled to change their eating, followed by the quip that all that needs to be done to fix the situation is to 'eat less and move more'. The implication is that those who do *not* eat less and move more are somehow lacking in moral fibre or brains. But consider this. American firefighters, who are not people notably lacking in courage or quick-wittedness, have higher rates of obesity and overweight – at 70 per cent – than the general population.[17] The way we eat is not a question of worthiness but of routine and preference, built over a lifespan. As the philosopher Caspar Hare has said: 'It is not so easy to acquire or drop preferences, at will.'[18]

Once we accept that eating is a learned behaviour, we see that the challenge is not to grasp information but to learn new habits. Governments keep trying to fix the obesity crisis with well-intentioned recommendations. But advice alone never taught a child to eat better ('I strongly advise you to finish that cabbage and follow it with a glass of milk!'), so it's strange that we think it will work on adults. The way you teach a child to eat well is through example, enthusiasm and patient exposure to good food. And when that fails, you lie. In Hungary, children are taught to enjoy eating carrots by being told that they bestow the ability to whistle. The point is that before you can become a carrot eater, the carrots have to be desirable.

When this book started taking shape in my head, I thought that my subject was childhood food. Bit by bit, I started to see that many of the joys and pitfalls of children's eating were still there for adults. As grown-ups, we may still reward ourselves with treats, just as our parents did, and continue to 'clean our plates', though they are no longer there to watch us. We still avoid what disgusts us, though we probably know better than to throw it under the table when no one is looking. Put a lit-up birthday cake in front of anyone and they are young again.

One of the questions I wanted to explore was the extent to which children are born with hard-wired preferences. As I trawled through endless academic papers in the library, I predicted fierce disagreement among contemporary scientists. On one side, I would find those who argued that food likes and dislikes were innate; and on the other, those who insisted they were acquired: nature versus nurture. To my astonishment, I found that this was not the case. Far from controversy, there was a near-universal consensus – from psychologists, from neuroscientists, from anthropologists and biologists – that our appetite for specific foods is learned.[19] Within this broad agreement, there are, as you might expect, still plenty of scholarly disputes, such as the brouhaha over whether our love–hate relationship with bitter vegetables such as Brussels sprouts has a genetic underpinning. There are also competing theories on the extent to which our food learning is mediated by particular genes, hormones and neurotransmitters. But the fundamental insight that human food habits are a learned behaviour is not the subject of scientific debate.

This scientific consensus is remarkable, given that it is the opposite of how we usually discuss eating habits in everyday conversation. There's a common assumption – shared, curiously enough, by those who are struggling to eat healthily and many of the nutritionists who are trying to get them to eat better – that we are doomed by our biology to be hooked on junk food. The usual story goes something like this: our brains evolved over thousands of years to seek out sweetness, because in the wild we would have needed a way to distinguish wholesome sweet fruits from bad bitter toxins. In today's world, where sugary food is abundant, or so the thinking goes, our biology makes us powerless to turn down these 'irresistible' foods. We know that tasting something sweet activates the pleasure-generating parts of the brain and even acts as an analgesic, comparable to drugs or alcohol. Palaeolithic brain + modern food = disaster.

What's missing from this account is the fact that, while the taste for sweetness is innate to all human beings and common to all cultures, when it comes to actual sweet foods – and other unhealthy processed foods – we show profoundly varied responses. As one 2012 study of food preferences states, our attitudes to sweetness vary 'in terms of perception, liking, wanting and intake'.[20] Different people enjoy sweetness in very different forms. Sweetness could mean a whole cob of corn at the height of summer; or a plate of milky-fresh mozzarella; or fennel cooked long and slow until it is toffee-brown. Our love of sweetness may be universal but there are vast individual differences in how we learn to ingest it. Put another way: not everyone wants to get their sweet hit in the form of Froot Loops.

Nutritionists use the word 'palatable' to describe foods high in sugar, salt and fat, as if it were impossible to prefer a platter of crunchy greens dressed with tahini sauce to a family-sized bar of chocolate. Yet around a third of the population – Palaeolithic brain or not – manage to navigate the modern food world just fine and select a balanced diet for themselves from what's available.

I'm not saying that to be thin is necessarily healthy. Some of the non-overweight may be anorexic or bulimic. Others avoid food through cigarettes and drugs or burn off a junk-food habit with manic exercise. When we speak of an 'obesity epidemic', as well as making those trying to lose weight feel worse than they already do, we miss the fact that the situation is more complex than thin = good and fat = bad. Professor Robert Lustig, a leading specialist on the effects of sugar on the human body, points out that up to 40 per cent of normal-weight people have exactly the same metabolic dysfunctions associated with obesity – 'diabetes, hypertension, lipid problems, cardiovascular disease . . . cancer and dementia' while around 20 per cent of obese people get none of these diseases and have a normal lifespan.[21]

So we cannot assume that everyone who is 'normal weight' has a healthy relationship with food. (Incidentally, given that these people are in a minority, isn't it time we stopped calling them 'normal'? How about 'exceptional' instead?) The situation is more complicated than the numbers suggest. But I'd still hazard that this exceptional third of the population has something important to tell us. There are hundreds of millions of individuals who somehow swim against the tide of the dysfunctional modern food supply and feed themselves pretty well. There are those who can eat an ice-cream cone on a hot day without needing to punish themselves for being 'naughty'; who automatically refuse a sandwich because it isn't lunchtime yet; who usually eat when they are hungry and stop when they are full; who feel that an evening meal without vegetables isn't really a meal. These individuals have learned the eating skills that can protect them in this environment of plenty.

Viewed through the lens of behavioural psychology, eating is a classic form of learned behaviour. There is a stimulus – an apple tart, let's say, glazed with apricot jam. And there is a response – your appetite for it. Finally, there is reinforcement – the sensory pleasure and feeling of fullness that eating the tart gives you. This reinforcement encourages you to seek out more apple tarts whenever you have the chance and – depending on just how great you feel after eating them – to choose them over other foods in the future. In lab conditions, rats can be trained to prefer a less sweet diet over a sweeter one when it is packed with more energy and therefore leaves them more satisfied: this is called post-ingestive conditioning.[22]

We know that a lot of this food-seeking learning is driven by dopamine, a neurotransmitter connected in the brain with motivation.[23] This is a hormone that is stimulated in the brain when your body does something rewarding, such as eating, kissing or sipping brandy. Dopamine is one of the chemical signals that passes information between neurons to tell your brain that you are having fun. Dopamine release is one of the mechanisms that 'stamps in' our flavour preferences and turns them into habits. Once animals have been trained to love certain foods, the dopamine response can be fired up in the brain just by the sight of them: monkeys have a dopamine response when they see the yellow skin of bananas as they anticipate the reward.[24] Anticipating dopamine release is the incentive that makes lab rats work hard for another treat by pressing a lever.

Humans, needless to say, are not lab rats.* In our lives, the stimulus–response behaviour around food is as infinitely complex as the social world in which we learn to eat. It's been calculated that by the time we reach our eighteenth birthday, we will have had 33,000 learning experiences with food (based on five meals or snacks a day).[25] Human behaviour is not just a clear-cut matter of cue and consequence, because human beings are not passive objects, but deeply social beings. Our conditioning is often indirect or vicarious. We do

---

* Strangely in humans, unlike in rats, obesity seems to be associated with reduced rather than heightened dopamine release, suggesting once more the complexity of our pleasure responses.

not just learn from the foods we put in our own mouths, but from what we see others eat, whether in our own families, at school or on TV.

As children watch and learn, they pick up many things about food besides how it will taste. A rodent can press a lever to get a sweet reward, but it takes an animal as strange and twisted as a human being to inject such emotions as guilt and shame into the business of eating. Before we take our first bite of a certain food, we may have rehearsed eating it in our minds many times. Our cues about when to eat and what to eat and how much extend beyond such drives as hunger and hormones into the territory of ritual (eggs for breakfast), culture (hotdogs at a baseball match) and religion (turkey at Christmas, lamb at Eid).

It soon became clear to me that I could not get the answers I sought about how we learn to eat without exploring our wider food environment, which is a matter of mealtimes and cuisine, parenting and gender as well as neuroscience.

Our modern food environment is fraught with contradictions. The burden of religious guilt that has been progressively lifted from our private lives has become ever more intense in the realm of eating. Like hypocritical temperance preachers, we demonize many of the things we consume most avidly, leaving us at odds with our own appetites. Numerous foods that were once reserved for celebrations – from meat to sweets – have become everyday commodities, meaning not only that we over-consume them but that they have lost much of their former sense of festal joy.[26] The idea that you don't eat between meals now seems as outdated as thinking you must wear a hat when you step out of the house.

Yet while the nutritional content of our food supply has

changed hugely over the past fifty or so years, other aspects of eating have not changed fast enough to keep pace with modern life. Parents are still using a range of traditional feeding methods – such as urging children to finish what's on their plate – that were devised for a situation where famine was always round the corner. As we'll see, such feeding techniques are directly contributing to child obesity, in cultures as diverse as China and Kuwait.

The theme I returned to more than any other was families. Most of what we learn about food happens as children, sitting at the kitchen table (if your family is lucky enough to have one), being fed. Every bite is a memory and the most powerful memories are the first ones. At this table, we are given both food and love, and we could be forgiven if, later in life, we have trouble distinguishing the two. It is here that we develop our passions and our disgusts and get a sense of whether it is more of a waste to leave something on the side of the plate or to eat it up when we are not hungry.

Our parents – like governments – hope we will learn about food from the things they tell us, but what we see and taste matters more than what we hear. In many ways, children are powerless at the table. They cannot control what is put in front of them, or where they sit, or whether they are spoken to kindly or harshly as they eat. Their one great power is the ability to reject or accept. One of the biggest things many children learn at that table is that our choice to eat or not eat unleashes deep emotions in the grown-ups close to us. We find that we can please our parents or drive them to rage, just by refusing dessert. And then the adults complain that *we* are difficult at mealtimes!

After a certain point in our lives, it is us and not our parents spooning food into our mouths. We discover the glorious liberation of being able to choose whatever we want to eat – budget permitting. But our tastes and our food choices are still formed by those early childhood experiences. Rather alarmingly, it seems that our food habits when we were two – whether we played with our food, how picky we were, the amount of fruit we ate – are a pretty accurate gauge of how we will eat when we are twenty.[27]

The acquisition of eating habits is a far more mysterious skill than other things we learn in childhood, such as tying our shoelaces, counting or riding a bike. We learn how to eat largely without noticing that this is what we are doing. Equally, we don't always notice when we have learned ways of eating that are dysfunctional, because they become such a familiar part of ourselves. Having particular tastes is one of the ways that we signal to other people that we are unusual and special. We become known as the person in the family who adores munching on bitter lemon rind or the one who eats apples right down to the pips.

You might say that food dislikes do not matter much: each to their own. I won't give you a hard time for hating the fuzzy skin of peaches if you will excuse my squeamishness about the gooey whites of soft-boiled eggs. The danger is when you grow up disliking entire food groups, leaving you unable to get the nutrition you need from your diet. Doctors working at the front line of child obesity say it has become common in the past couple of decades for many toddlers to eat no fruit and vegetables at all. This is one of the reasons constipation is now such a huge – though little mentioned – problem in Western countries, giving rise to 2.5 million doctor visits a year in the US.[28]

Some hold the view that it doesn't really matter if children have unhealthy tastes, because once they grow up they will effortlessly acquire a penchant for salad, along with a deeper voice and mature political opinions. Sometimes it does work out this way. Love and travel are both powerful spurs to change. In the 1970s it was a common rite of passage to reject the conventional bland watery foods of a 1950s childhood and embrace mung beans and spice. Many tastes – for green tea, say, or vodka – are acquired, if at all, in adulthood. When we learn to love these bitter but lovely substances, we undergo what psychologists call a 'hedonic shift' from pain to pleasure.[29] You may overcome your childish revulsion at the bitterness of espresso when you discover the wonderful after-effects, how it wakes up your whole body and infuses you with a desire for work. The great question is what it takes for us to undergo a similar 'hedonic shift' to enjoying a moderate diet of healthy food.

The process will be different for each of us, because all of us have learned our own particular way of eating. But wherever you start, the first step to eating better is to recognize that our tastes and habits are not fixed but changeable.

There's a danger here that I'm making the process of changing how you eat for the better sound easy. It isn't. In particular, it isn't easy for those who feed themselves on a tight budget. Many have observed that – in developed countries – obesity disproportionately affects those on low incomes. Poverty makes eating a healthy diet harder in numerous ways. It's not just because it is far more expensive, gram for gram, to buy fresh vegetables than it is to buy heavily processed carbohydrates. Maybe you live in a 'food desert' where nutritious ingredients are hard to come by; or in housing without an adequate kitchen. Growing up poor can engender a lifetime of unhealthy food habits, because a narrow diet in childhood is likely to narrow your food choices as an adult, even if your income later rises. When the flavour of white bread and processed meat are linked in your memory with the warmth and authority of a parent and the camaraderie of siblings, it can feel like a betrayal to stop eating them.

Yet it's striking that some children from low-income households eat much better than others, and sometimes better than children from more affluent families. The problems with how we eat now cut across boundaries of class and income. It is feasible to create decent, wholesome meals – bean goulash, spaghetti puttanesca – on a shoestring budget. Equally, one can have the funds to buy chanterelle mushrooms and turbot but no inclination to do so. According to feeding therapists with whom I have spoken, there are successful business people who will – literally – pass out from hunger at their desks rather than allow an unfamiliar meal to pass their lips when their preferred junk food is not available. Assuming you are not living in a state of famine, the greatest determinant of how well you eat is the way you have learned to behave around food.

This behaviour is often immensely complex. As we grow up, we become capable of second order preferences as well as first order preferences. A first order preference is basic: you love crispy roast potatoes, smothered in butter and salt. A second order preference is more convoluted: you *want to like* eating carrots instead of the potatoes because you think they would be less fattening and healthier. Indeed, you probably can, at least sometimes, limit yourself to eating raw vegetables instead of the carb-laden potatoes. But the real question is what happens next. In 1998 the social psychologist Roy Baumeister did a famous experiment. Baumeister, who is known for his work on self-defeating behaviours, found that the struggle of will required when a group of people were asked to eat 'virtuous' foods such as radishes instead of the foods that they really wanted, such as chocolate and cookies, led to diminishing returns.[30] They were so depleted by the effort of the task that when faced with another difficult task – solving a tricky puzzle – they would give up more quickly. The emotional effort of not eating the cookies had a 'psychic cost'.

Changing our food habits is one of the hardest things anyone can do, because the impulses governing our preferences are often hidden, even from ourselves. And yet adjusting what you eat is entirely possible. We do it all the time. Were this not the case, the food companies who launch new products each year would be wasting their money. After the fall of the Berlin Wall, housewives from East and West Germany tried each other's food products for the first time in decades. It didn't take long for those from the East to realize that they preferred Western yoghurt to their own.[31] Equally, those from the West discovered a liking for the honey and vanilla wafer biscuits of the East. From both sides of the wall, these German housewives showed a remarkable flexibility in their food preferences.

There is hope as well as concern in the fact that we remain like children in our eating patterns. We are like children in our fussiness and love of junk. But we also remain like children in that we have a capacity to learn new tricks, one that we seldom credit ourselves with. Even though most of us have tastes acquired very young, we can still change.

When I was a teenager I could eat whole pint-sized tubs of ice cream, and second and third helpings of everything. Everywhere I went, food screamed at me. Maybe it was a response to living with my older sister, who was anorexic, though this was never mentioned, because in our family we did not speak of such things. Or it could have been a consequence of growing up in a house where emotional talk was taboo. It definitely got worse when I was fourteen and my parents separated. Overeaters often say they are swallowing their feelings.

Around the age of twenty, something changed. I fell in love, got happier and my meals became more structured. I shrank from large to medium, without ever particularly dieting. I ate lots of vegetables, not because I had to, but because they were delicious, and they made me feel good. Then I had children. I could now bake a whole chocolate cake, eat a small slice and leave the rest. Recently I discovered yoga. My teenage self would have found my current self intensely annoying.

The strange thing, however, is that my behaviour changed without me ever particularly noticing that this was what was happening. Unlike the adolescent diets that I imposed on myself in a conscious, self-correcting way, this new healthier life crept up on me unawares. It's not that I never carry on eating crisps long after I am full, especially when there's a glass of wine in my hand. I may be safe around chocolate cake, but I wouldn't fancy my chances with a Vacherin Mont d'Or cheese in the kitchen. But I have definitely reached the point where my second order food preferences – I want to like greens – and my first order food preferences – I do like greens – are fairly in sync. Food no longer screams but speaks to me. It helps that our concept of healthy eating has enlarged in recent years to take in satisfying meals such as chicken and chickpea soup, buckwheat pancakes, avocado toast or buttery scrambled eggs with herbs. I'm in the groove now of eating smaller lunches and larger dinners, but small or large, meals are occasions for pleasure, not angst. This feels good. I must have relearned how to feed myself somewhere along the way, treating myself with some of the solicitude I bestow on my children.

E.P. Köster, a behavioural psychologist who has spent decades studying why we make the food choices we do, says that food habits 'can almost exclusively be changed by relearning through experience'.[32] That is, if we want to relearn how to eat, we need to become like children again. Bad food habits can only change by making 'healthy food' something pleasure-giving. If we experience healthy food as a coercion – as something requiring willpower – it can never taste delicious.

It's seldom easy to change habits, particularly those so bound up with memories of family and childhood, but, whatever our age, it looks as if eating well is a surprisingly teachable skill. This is not to say that everyone should end up with the same tastes. Life would be dull if everyone preferred satsumas to clementines. But there are certain broad aspects of eating that can be learned and then tailored to your own specific passions and needs. There are three big things we would all benefit from learning to do: to follow structured mealtimes; to respond to our own internal cues for hunger and fullness rather than relying on external cues such as portion size; and to make ourselves open to trying a variety of foods. All these three can be taught to children, which suggests that adults could learn them too.

For our diets to change, as well as educating ourselves about nutrition – and yes, teaching ourselves to cook – we need to relearn the food experiences that first shaped us. The change doesn't happen through rational argument. It is a form of reconditioning, meal by meal. You get to the point where not eating when you are not hungry – most of the time – is so instinctive and habitual it would feel odd to behave differently. Governments could do a great deal more to help us modify our eating habits. In place of all that advice, they could reshape the food environment in ways that would help us to learn better habits of our own accord. A few decades from now, the current laissez-faire attitudes to sugar – now present in 80 per cent of supermarket foods – may seem as reckless and strange as permitting cars without seatbelts or smoking on aeroplanes.[33] Given that our food choices are strongly determined by what's readily available, regulating the sale of unhealthy food would automatically make many people eat differently. Banishing fast-food outlets from hospitals and the streets surrounding schools would be a start. One study shows that you can reduce chocolate consumption almost to zero in a student cafeteria by requiring people to line up for it separately from their main course.[34]

But at an individual level, we won't achieve much by waiting for a world where chocolate is scarce. The question is what it might take to become part of that exceptional third (give or take) of the population who can live in the modern world, with all its sugary and salty allurements, and not be agonized or seduced. Having a healthy relationship with food can act like a life jacket, protecting you from the worst excesses of the obesogenic world we now inhabit. You see the greasy meatball sandwich and you no longer think it has much to say to you. This is not about being thin. It's about reaching a state where food is something that nourishes and makes us happy rather than sickening or tormenting us. It's about feeding ourselves as a good parent would: with love, with variety, but also with limits.

Changing the way you eat is far from simple, but nor, crucially, is it impossible. After all, as omnivores, we were not born knowing what to eat. We all had to learn it, every one of us, as children sitting expectantly, waiting to be fed.

CHAPTER 1:

Likes and Dislikes

Every man carries within him a world, which is composed of all that he has seen and loved, and to which he constantly returns, even when he is travelling through, and seems to be living in, some different world.

FRANÇOIS-RENÉ DE CHATEAUBRIAND,
Travels in Italy, 1828

'He won't eat anything but cornflakes,' complained the mother of a boy I used to know. Breakfast, lunch or dinner – always a bowl of cornflakes and milk. Even at other people's houses, this boy made no concessions. To his mother, his extreme diet was a source of worry and exasperation. To the rest of us, he was a fascinating case study. Secretly, I was slightly in awe of him; my sister and I would never have dared be so fussy. To look at, you wouldn't know there was anything different about this kid: scruffy blond hair, big grin, neither unduly skinny nor chubby. He was not socially withdrawn or difficult in any other way. Where did it come from, this bizarre cornflake fixation? It just seemed to be part of his personality, something no one could do anything about.

Whether you are a child or a parent, the question of 'likes and dislikes' is one of the great mysteries. Human tastes are astonishingly diverse, and can be mulishly stubborn. Even within the same family, likes vary dramatically from person to person. Some prefer the components of a meal to be served separate and unsullied, with nothing touching; others can only fully enjoy them when the flavours mingle in a pot. There is no such thing as a food that will please everyone. My oldest child – a contrarian – doesn't like chocolate; my youngest – a conformist – adores it. It's hard to say how much of this has to do with chocolate actually tasting different to each of them and how much it has to do with the social pay-off you get from being the person who either likes or loathes something so central to the surrounding culture. The one who loves chocolate gets the reward of enjoying something that almost everyone agrees is a treat. And he gets a lot of treats. The one who doesn't like chocolate gets fewer sweets, but what he does get is the thrill of surprising people with his oddball tastes. He fills the chocolate-shaped void with liquorice.

Yet my chocolate-hating boy will happily consume pieces of chocolate if they are buried in a cookie or melted in a mug of hot cocoa. One of the many puzzles about likes and dislikes is how they change depending on the context. As the psychologist Paul Rozin says: 'to say one likes lobster does not mean that one likes it for breakfast or smothered in whipped cream.'[1] Different meals, different times of day and different locations can all make the same food or drink seem either desirable or not. Call it the retsina effect: that resinated white wine that is so refreshing when sipped on a Greek island tastes of paint-stripper back home in the rain. It's also worth remembering that when we say we like this or that, though we use the same words, we are often not talking about the same thing. You may think you hate 'mango' because you have only ever tasted the fibrous, sour-yellow kind. When I say I adore it, I am thinking of a ripe Alphonso mango from India, brimming with orange juice and so fragrant you could bottle it and use it for perfume.

The foods we eat the most are not always the ones we like the most. In 1996 the psychologist Kent Berridge changed the way that many neuroscientists thought about eating when he introduced a distinction between 'wanting' (the motivation to eat something) and 'liking' (the pleasure that the food actually gives).[2] Berridge found that 'wanting' or craving was neurally as well as psychologically distinct from 'liking'. Whereas the zone of the brain that controls our motivation to eat stretches across the entire *nucleus accumbens*, the sections of the brain that give us pleasure when we eat occupy smaller 'hotspots' within this same area. For Berridge, this discovery offers a fruitful way for thinking about some of the 'disorders of desire' that bedevil humans. For example, binge eating may – like other addictive behaviours – be associated with 'excessive wanting without commensurate "liking"'.[3] You may feel a potent drive to purchase an extra-large portion of cheesy Nachos even though the pleasure they deliver when you actually consume them is much less potent than you expected. Indeed, binge eaters often report that the foods they crave do not even taste good when they are eating them: the desire is greater than the enjoyment.

However, several neuroscientists have pointed out in response to Berridge that liking and wanting remain 'highly entangled'.[4] Berridge himself admits that there is strong evidence that if you reduce the amount a food is liked, the consequence is that it is also wanted less.[5] Even if our craved foods do not make us as happy as we hope they will, the reason that we crave them in the first place is because we once loved them.[6] Like drug addicts, we are chasing a remembered high. Our 'likes' thus remain a central motivating force in shaping how and what we eat. To find out more about why we like the foods we do remains a vital question for anyone who is interested in feeding themselves or their family better. If asked to say where tastes come from, I suspect that most of us would say they were determined by individual temperament, which is another way of saying 'genes'. Being a chocolate lover – or hater – becomes so much part of our self-image that we can't imagine ourselves any other way. We show that we are adventurous by seeking out the hottest chillis; we prove we are easy-going by telling our host we 'eat anything'. We confirm that we are naturally conservative by eating patriotic hunks of red meat. Taste is identity. Aged eight, my daughter used to draw pictures of herself and write 'prawns-peas-mushrooms' at the top, surrounding herself with the tastes she loved best.

Because our tastes are such an intimate part of ourselves, it is easy to make the leap to thinking that they must be mostly genetic: something you just have to accept as your lot in life. Parents often tell children that their particular passions place them on this or that side of the family – you got your fussiness from your grandfather! – as if you were destined from birth to eat a certain way. Sometimes it is uncanny how a suspicion of celery or a deep hunger for blackberries replicates from parent to child. When we notice these familial patterns, it confirms us in our view that food preferences must be inherited through our genes.

When I've described the argument of this book to people I meet, sometimes they get a little angry. 'I disagree that we learn how to eat,' they say. 'You'd never get me to like sultanas/squid/salami [delete as appropriate].' Anyway, they say, 'What about genes?'

It's fine by me if you don't like sultanas. And I'm certainly not denying that there is a genetic component in our relationship with food. We are not born as blank slates. Some people have a heightened genetic sensitivity to certain flavours (notably bitterness) while others are blind to them.[7] There are also genetic variations in individual appetite, the speed at which we eat and the extent to which people actually enjoy eating.[8] We vary in how we chew, how we swallow and how we digest. Some people are born with conditions that make it much harder to eat, such as a delay to the oral-motor system. I had no idea quite how fraught the basic matter of getting food from plate to mouth could be until my third child was born with cleft palate and he and I both struggled at mealtimes. He is now five and new dishes occasionally still provoke tears (usually his). Our relationship with food and weight is additionally affected by epigenetics: our experience in the womb. The 'thrifty phenotype' hypothesis of biochemist C. Nicholas Hales and epidemiologist David Barker suggests that being undernourished in utero leaves people with a lifelong propensity for weight gain, an unfair fate to be handed so early.[9]

The question remains to what extent we are capable of overriding this genetic and epigenetic inheritance and learning new tastes. This riddle can seem impossible to unravel, given that children do not learn to eat under laboratory conditions. As we take our first bites, our parents are supplying us simultaneously with both nature (genes) and nurture (environment conceived in its broadest sense, including every-thing from cuisine to family dynamics to religion to cutlery and table manners to the ethics of meat to views on whether it's OK to eat food off the floor if it was only there for five seconds). The two are so intertwined, it's hard to tell where one starts and the other stops.

In one remarkable experiment, however, a group of chil-dren did learn to eat under lab conditions. In the 1920s and 1930s, Dr Clara Davis, a paediatrician from Chicago, spent six years trying to study what children's appetites would look like if allowed to blossom in total freedom without any pre-conceived ideas of what tasted good.[10] Davis's results have often been taken as a clear indication that likes and dislikes are fundamentally inbuilt and natural though, as we'll see, Davis herself drew a rather different conclusion.

In 1926, at Mt Sinai Hospital in Cleveland, Dr Clara Marie Davis started the most influential experiment ever conducted into the question of human likes and dislikes. As a doctor, Davis saw many children with eating problems – mostly refusal to eat – whose appetites did not match their nutritional needs. She wondered what children's appetites would look like, freed from the usual pressures of parents and doctors pushing them to eat nutritious foods such as hot cereal and milk, regardless of whether they liked them. Conventional medical wisdom at that time was that children's particular likes should not be indulged, lest they became 'faddy'. Dr Davis was not so sure that eating what you liked was auto-matically a bad thing.

She borrowed a number of infants – some of them orphans from institutions and some the children of teenage mothers and widows – and placed them on a special 'self-selection diet' under her medical care. The children – aged from six to eleven months, who had never yet tasted solid food – were offered a selection of whole, natural foods and given free rein, day after day, to eat only what they wished. The full list of foods was:

1. Water	19. Lettuce
2. Sweet milk	20. Oatmeal
3. Sour (lactic) milk	21. Wheat
4. Sea salt	22. Corn meal
5. Apples	23. Barley
6. Bananas	24. Ry-Krisp
7. Orange juice	25. Beef
8. Fresh pineapple	26. Lamb
9. Peaches	27. Bone marrow
10. Tomatoes	28. Bone jelly
11. Beets	29. Chicken
12. Carrots	30. Sweetbreads
13. Peas	(offal from the pancreas)
14. Turnips	31. Brains
15. Cauliflower	32. Liver
16. Cabbage	33. Kidneys
17. Spinach	34. Fish (haddock)[11]
18. Potatoes	

At each meal, the infants were offered a selection of around ten foods from this list, all of them mashed, ground up or finely minced. Some, such as bone marrow, beef, peas and carrots, were offered both in cooked and raw form. The selection was laid out in bowls, while nurses sat by, waiting to see what the children would choose. As Davis described it:

The nurse's orders were to sit quietly by, spoon in hand, and make no motion. When, and only when, the infant reached for or pointed to a dish might she take up a spoonful and, if he opened his mouth for it, put it in. She might not comment on what he took or did not take, point to or in any way attract his attention to any food, or refuse him any for which he reached. He might eat with his fingers or in any way he could without comment or correction of his manners.[12]

Davis continued this experiment over a period of six years, starting with three babies and building up to fifteen. The results, which have been hotly discussed by doctors ever since, were dramatic. Without any preconceived notions about what foods were suitable for them, the babies showed enthusiasm for everything from bone marrow to turnips. They didn't realize they weren't supposed to like beets or offal. All of them tried all of the thirty-four foods, except for two who never attempted lettuce and one who shunned spinach.

Within a few days, Davis noticed, 'they began to reach eagerly for some and to neglect others, so that definite tastes grew under our eyes.'[13] It soon became obvious to her that for the fifteen children, there were '*fifteen different patterns of taste*'. The children made some very odd selections which looked like a 'dietician's nightmare' said Davis. They went on curious 'food jags'. One day, they might gorge on liver or eat a meal of nothing but bananas, eggs and milk. A boy called Donald showed a rare passion for oranges, cramming in nearly two pounds of them one day.[14] In the process of trial and error of finding out what tasted nice, some of the children 'chewed hopefully' on plates and spoons, while others grabbed handfuls of pure salt. On trying something new, Davis observed that their faces showed first surprise, then indifference, pleasure or dislike.

However bizarre and unbalanced the children's likes and dislikes look to our eyes, they served them well. In a 1928 article writing up her findings, Dr Davis included a 'before' and 'after' photo of one of the children, Abraham G. At eight months, on arriving in her care, he looks a little pale. At twenty months, after a year on the diet, he is cherubic and plump.

When they arrived at the hospital the infants were generally in poor health. Four were seriously underweight; five had rickets. Yet within a few months, all the children were pink-cheeked and optimally nourished. One of the rickets sufferers was offered cod liver oil, which he took the occasional glug of; but the other four managed to get enough vitamin D and calcium to cure their rickets through diet alone. When they suffered colds, they appeared to self-medicate, eating vast amounts of carrots, beets and raw beef. Even though they were given no guidance on what their bodies needed, their ratio of calories averaged at protein 17 per cent, fat 35 per cent and carbohydrate 48 per cent, very much in line with contemporary nutritional science.

Dr Davis created an unprecedented body of information on childish appetites (though it was never fully analysed and, after her death in 1959, all the boxes of raw data were discarded). When Davis took up a new job, the original set-up in Cleveland was moved to Chicago, where she established what amounted to 'an eating-experiment orphanage'. In all, she logged around 36,000 meals as well as recording changes in height and weight, blood and urine, bowel movements and bone density.[15] It is unlikely any scientist will ever get such

detailed data again, given the dubious ethics of keeping children locked up in an experimental nursery for so long. The
babies stayed on the diet for a minimum of six months and
a maximum of four and a half years, during which time they
were always at the hospital.

No friends visited and those who were not orphans had
little or no contact with their parents. While in the hospital
nursery, their lives were subordinated to the needs of the
experiment. Such an arrangement would never be allowed
now, though Davis evidently cared for the children very much,
in her way. She adopted two of them, as a single mother:
Abraham G (the plump cherub) and Donald, the passion-
ate orange eater. Many years later, after Donald was dead,
his widow recalled that he and Abraham had always been
'easy to cook for' and 'happy to try all kinds of foods'; they
remained omnivores all their lives.[16]

It was such an extraordinary, audacious, borderline-crazy
project that Davis attempted: to get to the heart of where
children's food passions come from. It's just a shame that
her experiment proved so easy to misread. Time and again,
Davis's orphanage has been held up as evidence that appetite is mostly genetic and, as a consequence, that the foods
children like or dislike are a sure guide to what their bodies
need.[17] Davis's food orphanage has been taken as proof that
in their natural state, likes and dislikes are genetic and highly
individual, like fingerprints: our tastes are a matter of nature,
not nurture. What this interpretation fails to take account of
is that the biggest thing Dr Davis did was to radically restruc-
ture the food environment of the children.

There was a 'trick' to the way the experiment was set
up, as Clara Davis was the first to point out. The real secret
was in her choice of the thirty-four foods, which were all

unprocessed whole foods. With such foods preselected for them, it didn't matter which ones the children were drawn to on any given day, because, assuming they took food from several of the bowls at each meal, they could not help but eat a diet of an excellent standard of nutrition. Davis said that her choice of food was designed to mimic the conditions of 'primitive peoples', though the heaping bowlfuls were surely more plentiful than any hunter-gatherer regime. The experiment proved that when your only food choices are good ones, preferences become unimportant. The 'fifteen patterns of taste' resulted in a single healthy whole-food diet, because of the set-up. None of the children was totally omnivorous, but nor were their likes and dislikes a problem, as they so often are in normal family life. There was no option to like unhealthy food and dislike healthy food.

Davis herself concluded that her experiment showed that the selection of food for young children should be left 'in the hands of their elders where everyone has always known it belongs'. Instead of the 'wisdom of the body' Davis spoke of the 'glaring fallibility of appetite'. It was obvious to her that there was no 'instinct' pointing blindly to the 'good' and the 'bad' in food. The two most popular foods overall in her study were also the sweetest: milk and fruit. Had she offered the children a free choice of 'sugar and white flour', those staples of a 1930s diet, it is unlikely they would have ended up in such fine fettle. Self-selection, she concluded, would have little or no value if the children were selecting from 'inferior foods'.

The real test, Davis recognized, would be to offer newly weaned infants a choice between natural foods and processed food. This was to have been her next experiment, but the Depression dashed this prospect, as her funding ran out at the crucial moment. Davis never got the chance to test the effects on appetite of the 'pastries, preserves, gravies, white bread, sugar and canned food' that had in her lifetime become so popular. Davis's experiment left a powerful legacy that took no account of the trick at the heart of it. Doctors, particularly in America, interpreted her experiment to mean that children's appetites are inbuilt and benign, without paying attention to the way in which Davis had changed the food environment in which the babies ate. Her work was seized on as proof that our individual appetites are messages encoded with exactly the nutrients that our particular body needs. If we need protein, we will crave chicken. If we have rickets, we will naturally gorge on vitamin D until we are cured. All we have to do to eat well is listen to our cravings. Mother Nature knows best. Davis herself gave licence to such a view, commenting that the children's successful 'juggling and balancing' of more than thirty essential nutrients suggested 'the existence of some innate, automatic mechanism . . . of which appetite is a part'.[18]

Influenced by Davis, the dominant view on appetite among paediatricians became 'the wisdom of the body', which went along with the vogue for 'child-centred' learning. In 2005 Benjamin Scheindlin MD, a paediatrician, noted that Davis's work contributed to a widespread change in attitudes in paediatric medicine from the 1930s onwards.[19] Where a previous generation lamented the pickiness of children's changeable tastes, now doctors positively welcomed childish vagaries of appetite. Dr Spock, author of the bestselling *Baby and Child Care*, first published in 1946, devoted ten pages to the Davis experiment. A mother, in Spock's opinion 'can trust an unspoiled child's appetite to choose a wholesome diet if she serves him a reasonable variety and balance'.[20] It didn't matter if a child developed a temporary dislike of a vegetable, because their cravings would naturally give them everything they needed in the way of nutrition.

Many experts in child-rearing still think like this, operating on the assumption that children are born with special appetites for exactly the nutrients they most need and that it will all balance out, if only they are given free rein to eat what they like. A book on solving children's eating problems that went through several reprints in the 1980s and 1990s argued that the implications of Davis's work were that children should be given total control over food selection: let them eat cornflakes![21] As recently as 2007, a popular website about feeding children discussed Davis and concluded that there was 'a strong biological plausibility . . . that children will instinctively choose a balanced diet'.[22]

The 'wisdom of the body' is an alluring thought (like maternal instinct and other biological myths). Eating would be such a simple business, if only we had little memos inside our bodies telling us what we needed to eat at each precise moment (your vitamin C levels are dropping – quick, eat a kiwifruit!). If only we liked just the stuff that was good for us and disliked anything superfluous or bad. We can certainly learn to get better at reading our body's cues for food, but this tends to come with age and experience, as you notice little things like how pasta for lunch makes you sleepy or that a handful of nuts and a cup of Greek yoghurt keep you full for longer than white toast and jam. But children's omnivorous bodies – after the milk stage, when breastfed infants do self-regulate – are not so wise.

Many children habitually seek out precisely the foods that are least suitable for them. They crave sugar and shun green vegetables. They neglect to drink enough water. Nutritious meals are rejected, while junk is revered. Can we really believe that a preschooler demanding a packet of the latest kids' sugary breakfast cereal, having seen it on TV, is responding to their body's need for certain vitamins and carbohydrate?

The scientific evidence – both from humans and rats – shows that the theory of the 'wisdom of the body' is flawed at best. For the theory to be true, omnivores would need to have specific appetites for the essential nutrients the body needed at any given time. This is a very unlikely proposition, given that the list of nutrients needed by omnivores comes in so many guises, depending on the environment we happen to live in. An innate appetite for the vitamin C in blackcurrants would be no use if you lived somewhere that blackcurrants do not grow. In lab conditions, rats – our fellow omnivores – have shown a very erratic ability to self-select the diet that would do them the most nutritional good. In one study, a group of rats were given a choice between a bad-tasting but protein-rich diet and a good-tasting but low-protein diet. Over the course of a week, fourteen out of eighteen rats failed to develop a preference for the food that would have done them the most good and they lost weight.[23] Other trials have attempted to find out whether rats could 'self-select' to correct certain vitamin deficiencies; and concluded that many of them could not. With thiamine-deprived rats, the process of learning to like a thiamine-rich diet took a week or more and the rats who did not adapt quickly enough to the correct food died.[24] As for human subjects, there is, notes one specialist in the field, no data to suggest innate appetites for specific foods. It does seem possible for humans to learn over time specific appetites that will correct certain imbalances – particularly a craving for salt when lacking in sodium – but that is a different matter.[25]

Ninety years after Davis's experiment, the view that likes are predominantly innate – or genetic – looks shaky. When trying to get to the bottom of where tastes come from, scientists have often turned to twins. If identical twins share more food likes than non-identical twins, the chances are that there is a genetic cause. Twin studies suggest that many aspects of eating are indeed somewhat heritable. Body weight – measured as BMI – appears to be highly heritable in both boys and girls.[26] So is dietary 'restraint', or the mysterious urge to resist eating the thing you want to eat.[27] But studies that look at likes and dislikes are much less conclusive. Several twin studies have suggested that identical twins are more likely to enjoy the same protein foods than non-identical twins, but when it came to snacks, dairy and starchy foods, their likes were only marginally more similar than those of the non-identical twins.[28] Overall, the evidence for tastes being heritable is very modest, accounting for only around 20 per cent – at most – of the variation in foods eaten.[29]

Genes are only ever part of the explanation for what we choose to eat. As one senior doctor working with obese children put it to me, you could be cursed with all the genes that make a person susceptible to heart disease and obesity and still grow up healthy, by establishing balanced food habits. 'All of it is reversible,' he said. Parents and children resemble each other no more in the foods they like than couples do, suggesting that nurture – who you eat with – is more powerful than nature in determining our food habits.[30] Whatever our innate dispositions, our experience with food can override them. Maybe the reason you share your parent's hatred of celery is because you have seen them recoil from it at the dinner table.[31] Researchers found that when they gave three groups of preschool children different varieties of tofu – one group had plain tofu, one ate it with sugar and one with salt – they quickly came to prefer whichever one they had been exposed to, regardless of their genes.[32] It turns out that, so far from being born with genetically predetermined tastes, our responses to food are remarkably open to influence, and remain so throughout our lives.

If you want to know what foods a person does and does not like, the single most important question you can ask is not 'What are your genes?' but rather 'Where are you from?'[33] Had he lived in a part of the world where cornflakes are hard to come by, the cornflake boy would have had to find another way to annoy his parents. To a large extent, children eat – and therefore like – what's in front of them, particularly in conditions of scarcity. 'If you want your children to be less fussy about what they eat,' a friend who had fallen on hard times during the recession advised me, 'I can recommend poverty.' It's not really an option to be picky about the staple food of rice if you live in rural China.

Genes do make a difference – to the foods we like, the way we taste them and even how much we enjoy eating – but they turn out to be much less significant than the environment we learn to eat in. Contrary to our deepest beliefs about ourselves and our children, our likes and dislikes – the important ones, anyway, such as whether we eat enough vegetables or how much variety and balance we have in our diets – are much more about nurture than nature. Apart from changing the infants' food environment, there was another bigger trick to Davis's experiment, which she did not mention, perhaps because it is so obvious. She radically changed their social experience when eating, removing all extraneous social influence. In place of the hubbub of the family dinner table, the babies had only expressionless nurses who 'might not comment' in any way on their choices. The thought of being served in this silent, impassive way is creepy, particularly for the oldest children, who must have been as old as five by the time they left the orphanage. They ate without anyone caring what they ate; without any siblings fighting them for the last slice of pineapple; without any surrounding ideas about cuisine.

Davis was mistaken if she thought this was the way to discover the true nature of children's appetites. Though the nutritional outcomes were excellent, it was a not-quite-human way to eat, and one which no child in a real situation will ever replicate. We cannot arrive at the truth about appetite by removing all social influences. Appetite is a profoundly social impulse. To a large extent, our likes and dislikes are a response to the environment we eat in. From our first toothless tastes, we are picking up cues about which foods are desirable, and which are disgusting, which sadly are so often the very ones the grown-ups most want us to eat.

The public discussion of eating habits is focused on temptation and the idea of resisting desirable foods. But if we look at eating through the eyes of a child, we see that disgust may be even more powerful than desire in forming our tastes. Our urge to avoid eating something that makes us feel sick is often at the root of disordered eating, as we swerve away from whole categories of foods that we imagine would make us feel queasy. The most common reason for disgust is nausea: anything eaten just before a bout of stomach bug may be hated for life. Psychologist Paul Rozin, the world's leading expert on disgust, has argued that a central feature of disgust is 'contagion: when a disgusting food touches otherwise acceptable foods, it renders them permanently inedible'.[34] And yet most of the foods that we happen to find disgusting are not toxins but perfectly edible and wholesome foods. Brussels sprouts, for example.

If there is one food associated with personal dislikes in the Anglo-Saxon world, it is the Brussels sprout. Many people assume they have no choice in this matter – they just can't stand them. Are they right? In an article singing the praises of Brussels sprouts, the great chef Yotam Ottolenghi noted that there was a 'genetic explanation for why people either love or loathe' these little green brassicas.[35] Ottolenghi argued that being a sprout hater was likely to be a consequence of having a certain gene – TAS2R38 – which 'makes a protein that reacts with a chemical called PTC to create the sensation of bitterness'. Could this really be true? Is there a molecular basis to our hatred – or otherwise – of green vegetables?[36]

Some people definitely taste certain flavours more acutely than others. To take one of the stranger examples, up to 30 per cent of the population cannot physically pick up on androstenone, one of the key aromas that make truffles such a luxury. If you served them a sumptuous plate of pappardelle with truffle shavings, they would have no idea why it was meant to cause such joy. A different minority have a heightened sensitivity to coriander leaf, making it taste soapy and gross, rather than herbal and fresh. And, as Ottolenghi says, we vary hugely in our response to bitter tastes. All babies find bitterness somewhat horrible, which is probably a survival mechanism, given that in the wild, toxic substances tend to be bitter. The bitter response of a newborn includes arched lips, a protruding tongue, an expression of anger and spitting: all pretty vivid signs that babies do not consider bitterness to be yummy. Over time, however, it is possible to learn to love bitter substances: witness the fact that the world's two most popular beverages are coffee and beer.

Some learn to love bitterness; some tolerate it because they enjoy the buzz they get from a bottle of IPA or a cup of strong cafetière coffee; and some hardly taste it at all. Linda Bartoshuk of Yale University was the first to use the term 'supertaster' in the mid-1990s to refer to individuals with a heightened response to certain tastes, predominantly bitter ones (the phenomenon was first observed in the 1930s). Bartoshuk and colleagues found that there were significant genetic differences in the way we perceive bitterness. PROP (6-n-propylthiouracil) and PTC (phenylthioucarbamide) are chemical substances that either taste incredibly bitter or slightly bitter or of nothing at all, depending on whether you have the gene to taste them.[37] Around half of us are medium tasters, a quarter are non-tasters and another quarter are supertasters. Women are more likely to be supertasters than men. Bartoshuk has shown that PROP supertasters have more taste buds on their tongue than non-tasters. There's a very simple way to self-diagnose whether you are a super-taster or not. Swab your tongue with a little blue food dye and place a hole punch reinforcer ring on your tongue. Count how many pink bumps you can see inside the ring – these are the fungiform papillae, each containing 3–5 taste buds. If fewer than 15, you are a non-taster. If 15–35 you are a medium taster. If more than 35, you are a supertaster.

Psychologists got excited about the concept of PROP tasting, because it seemed to hold out – at last – the genetic key to likes and dislikes. Could bitter sensitivity be the secret of why some people eat unhealthy diets with few or no vege-tables? Is it because they lack a gene for sprouts? The world of flavour must be a very different place to PROP supertasters and non-tasters and it would appear obvious that this would translate into food habits. When seventy-one women and thirty-nine men were asked to taste asparagus, kale and Brussels sprouts, the PROP supertasters did indeed find the vegetables to be more bitter and less sweet.[38]

The surprising thing, however, is that, from a mass of research into PROP tasting, very little does point to genes determining food choices, either in children or adults.[39] Over time, your PROP status is not a particularly strong predictor of what your likes and dislikes will be. If anything, PROP non-tasters – the ones who can't taste bitterness in the sprouts at all – are slightly more at risk of an unhealthy diet and weight than the PROP supertasters.

There's clear evidence that PROP supertasters are more sensitive to certain flavours: the burn of chilli, the warmth of cinnamon, the acrid glow of coffee, the rasp of alcohol, the aftertaste of sweeteners and grapefruit – all these are perceived more strongly, often unpleasantly so. What is not so predictable is how this affects preferences. Given that supertasters perceive alcoholic drinks as more bitter, you'd expect them to drink less of them – indeed being a non-taster has been identified in some studies (though not others) as a risk factor for alcoholism: if whisky tasted like water, how easily it might go down. But a study of young adults found that being a PROP taster did not predict how much beer was drunk. After decades of enjoying countless glasses of wine from all the great *terroirs* in the world, the leading wine writer Jancis Robinson found out that she was a supertaster, something that in theory should make wine taste odiously acrid to her. That's not how it turned out. As she put it: 'If I enjoy wine less than the rest of you, you are very lucky wine drinkers indeed.'[40]

When it comes to childhood, the key question is whether being a PROP taster sets you up for a lifetime of disliking the leafy green vegetables every nutritionist wants us to eat more of. Greens – especially those in the cabbage family – contain bitter-tasting glucosinolate compounds. One study suggested that PROP-tasting children were more likely to dislike raw broccoli, but not cooked broccoli. Another study found that when offered black olives, cucumber and raw broccoli, PROP non-taster children ate a larger quantity than tasters did.[41] But when studies have looked at actual preferences rather than what children are prepared to eat in front of researchers, the signs are that PROP tasting in no way dooms you to dislike bitter vegetables. When 525 Irish children (aged seven to thirteen) were asked to record their intake and liking of cabbage, cauliflower, Brussels sprouts and broccoli over a three-day period, there were few significant differences between tasters and non-tasters.[42] The supertasters did show a marginally lower liking for Brussels sprouts and non-tasters liked cauliflower the most. But when their consumption of bitter vegetables overall was totalled up and averaged out, there were no differences in intake for PROP tasters and non-tasters. In this study, being a PROP taster mattered less than the simple fact of whether these Irish children were boys or girls: girls tended to like bitter vegetables more, or at least to be polite enough to pretend that they did.

A 2013 survey of college students pointed to a similar conclusion. The supertasters and the non-tasters showed no marked difference in likes and dislikes for: Brussels sprouts, broccoli, cabbage, spinach, crushed red pepper, jalapeño peppers, red wine, beer, salad dressing, mayonnaise.[43] The only substances that emerged as having significant negative connotations for PROP tasters were dark chocolate, coffee and chilli: the dark pungent end of the bitter flavour spectrum. The team of researchers concluded that environment mattered more than genes in determining preference. In America, they noted, many people 'know they are not going to like spinach, tofu, liver or "healthy food" and learn that fast food burgers, soda pop and sweet breakfast cereals are delicious . . . before they ever take a bite'.[44]

Some of the most telling research to date on PROP tasters looked at how genes interacted with the food environment children were growing up in and confirmed that household income and access to good food are more critical in forming tastes than being a supertaster. Over five years from 2005 to 2010, researchers studied 120 New York children aged four to six. Their PROP status was measured and they were deemed to be living in either a 'healthy food environment' or an 'unhealthy food environment', as judged by the slightly crude method of dividing the number of healthy food sellers by the number of unhealthy food sellers within a half-mile radius of where they lived.[45] In a healthy food environment, likes and dislikes followed the pattern that Ottolenghi – and common sense – would suggest. In this experiment, unlike the Irish one mentioned above, the PROP non-taster children who couldn't detect bitterness did indeed show a higher acceptance of vegetables – with fewer dislikes – than taster children in the same healthy environment. The interesting – and troubling – result was what happened to the children in the unhealthy food environment. Here, the likes and dislikes of tasters and non-tasters were not very different. The big difference was in the BMI of the children. In the unhealthy environment, the non-taster children had a higher BMI than any of the other groups studied. Their average BMI was over 1.6, which counts as obese.

What matters most for determining whether your tastes will be healthy ones is not whether you have a sprout-hating gene but the way that your genetic predispositions interact with your food environment. Once environment is taken into account, being a non-taster poses bigger health risks in our current state of plenty and junk than being a supertaster. Several studies have now found that non-tasters – adults as well as children – are the ones who tend to have higher BMIs. The theory is that non-tasters – since they do not experience certain flavours with the same intensity – are more responsive to the influences around them, for good or ill. They learn their likes more easily than supertasters. In a healthy food environment, they will easily acquire healthy tastes. When offered vegetables, they are less likely than supertasters to dismiss them as too bitter. But if they learn to love the wrong foods, the non-tasters can find themselves – like those New York children – obese by the age of six.

So, no, you can't blame your dislike of sprouts simply on having a faulty gene. If everyone's first nibble of sprouts was of Ottolenghi's own sprouts with caramelized garlic and lemon peel, charred in a hot pan until sweetly blackened at the edges, maybe they would be the most popular of all the vegetables. Perhaps your parents were sprout haters and – without meaning to – turned you against them. Or perhaps they forced them on you too vehemently. I know someone – a PROP supertaster, as it happens – who says she can never enjoy Brussels sprouts – though she has no quarrel with broccoli – because of memories of Christmas Day, when she was compelled by her parents to cut each hated sprout into quarters, and swallow them unchewed, like bitter pills. Maybe you never actually tasted sprouts because you 'knew' you wouldn't like them, because in our society the child who loves sprouts is considered a little odd. When the food writer Michele Humes arrived in the US from Hong Kong, it took her a while to get her head around the concept that 'children weren't supposed to like vegetables'.[46]

Likes and dislikes cannot be reduced to molecules and genes. This is bad news for the more sensationalist health pages, which thrive on headlines like 'Revealed: the Obesity Gene'. For the rest of us, it is – potentially – excellent information. It means that our food habits are not final and fixed but adaptable and open, if only we will give ourselves half a chance. We did not come into the world disliking bitter greens; we were taught to dislike them by our environment. Taste may be identity but it is not destiny. The hope – and admittedly it's a slim one at present for the children whose dislikes are vegetables and whose likes are all junk – is that while we are stuck with our genes, the environment is something that can change.

The main way we learn to like foods is simply by trying them. The term 'mere exposure' was coined by Robert Zajonc in 1968.[47] Zajonc's thesis was that affection is triggered by familiarity; and that disliking, conversely, is fear of the novel. Some of Zajonc's early experiments involved showing subjects complex shapes for very short periods of time. When the subjects were later asked to choose their favourite shapes from a line-up, there was a marked preference for the shapes that they had already encountered. Zajonc has suggested that there are similar forces at work when we favour Brie over Camembert.[48] These cravings are a function of prior experience. One or other cheese may trigger a recognition in us that we cannot necessarily put into words. Zajonc later observed this phenomenon of 'mere exposure' at work across cultures and species.

It's a truism that we know what we like and we like what we know. If you ask young children which foods they most detest, they tend to be the ones they have never actually tasted, often vegetables. To an adult, this sounds crazy: you can't know if you hate something until you have tasted it. 'Go on – you might like it!' I find myself urging, ineffectually, at the dinner table. But to a child, there is nothing paradoxical in saying, 'I don't like it – I never tried it!'[49] The foods that ranked highly on the 'never tried' list of a group of seventy American eight-year-olds included avocado (49/70 had never tried it), beetroot (48), prunes (43), collard greens (49), rye bread (43), lima beans (39), radish (38) and fried liver (55).[50]

The children's book *Bread and Jam for Frances* by Russell Hoban is about precisely this dilemma. Frances – a young badger – does not want to eat anything except bread and jam. 'How do you know what you'll like if you won't even try it?' asks her father. Eventually, her parents give in to her demands for nothing but bread and jam. She is delighted. But over time, being excluded from what the rest of the family is eating makes her sad and she craves variety. One evening, Frances begs tearfully for some spaghetti and meatballs. Her parents express surprise, because they didn't think she liked spaghetti. 'How do you know what I'll like if you won't even try me?' is her reply.

If liking is a consequence of familiarity, it follows that children are bound to like a narrower range of foods at first than adults, because they haven't tried as many. Problems arise when parents interpret this temporary wariness as something permanent. This is an easy mistake to make. The key period for acquiring preferences is toddlerdom: from one to three. But this coincides with a period in the child's life when they are most maddeningly, wilfully reluctant to try anything new. All children suffer from neophobia to a greater or lesser extent – a fear of new foods, often novel vegetables but also very commonly protein foods such as fish and meat. This reaches a peak between two and six. It probably evolved as a safety mechanism to protect us from toxins as we foraged in the wild. Now, unfortunately, it leads children away from the very foods they need to learn to like – vegetables and protein – and towards the comforting embrace of cakes, white bread and doughnuts.

As the name suggests, neophobia isn't just a dislike of how something tastes: it is an active fear of tasting it. In many cases, neophobia can be broken down simply by feeding the food to the child numerous times – often as many as fifteen – until the child realizes they haven't suffered any adverse consequences. See, the tomato didn't kill you! See, it didn't kill you again! Bit by bit dislike is lessened until one day it flips, almost comically, to enthusiasm. This has to be done over and over for each new ingredient. A child's love of cantaloupe is no guarantee that they will like watermelon.

The biggest problem with using 'mere exposure' on children is that you first have to persuade them to try the food. Exposing a child to broccoli multiple times is easier said than done. As any parent who has ever tried to feed a recalcitrant toddler will know, the best-intentioned strategies often backfire. 'Eat your vegetables and you can have a sweet' is a dangerous game to play because it makes the child dislike the vegetables even more. Psychologists call this the over-justification effect.[51] When a reward is offered for performing an activity, that activity is valued less. The child ends up loving sweets more, because they have become a prize.

Given that neophobia is a deep-seated fear that the unfamiliar food will cause you harm, it can help if the child
witnesses someone else eating the food and surviving; preferably even enjoying it. I did not know that this was what I was
doing, but after various futile attempts to get my daughter,
then three, to eat something green other than cucumber, I
hatched the idea of bringing her favourite doll to eat with
us. This doll – a grubby-faced baby boy – sat at the table and
proceeded to 'eat' green beans, as he oohed and ahhed with
ecstasy (or rather, I did). It felt pretty lame, but one day my
daughter begged to be given some of the baby doll's green
beans too and has loved them ever since. Another successful
strategy is combining a scary new food with a familiar old
one. Both children and adults are more likely to try something
new when it is served with a familiar condiment – a blanket
of ketchup, say, that renders the new food safe enough to try.
But as the food psychologist John Prescott has written, no
amount of ketchup will induce most children to try a plateful
of spiders.[52]

Most children get over the worst of their fear of new
food by the age of six or seven. Up to this age, it is considered a normal stage of child development. Having conquered neophobia, they may flip over to neophilia: an ostentatious delight in novel flavours that can look suspiciously
like showing off. My oldest child, the one who doesn't like
chocolate, is like this. His favourite foods change with capri-
cious haste; dishes may please him at first, then bore him.
He abhors plainness, grumbling that I always cook the same
things for supper (charming!) and taking a macho delight in
strongly flavoured condiments. When he was eight, we went
to Rome, just the two of us. At a famous offal restaurant, he
selected from the menu a dish called 'artichokes with lamb's

hearts and all the organs in the vicinity'. And ate it too, with gusto.

For a significant minority, however, a terror of new food – or mixed-up food, or strange food or spicy food or food that just plain smells wrong – is never conquered. The numbers are high: it has been estimated that as many as a quarter of all adults are severely neophobic about what they eat. Fussiness in children is something we often joke about or laugh off. The cornflake boy was seen – outside his family, anyway – as a comic figure rather than a tragic one.

But living as a neophobic adult is no joke. I've met grown men and women who quietly confessed that they could not bring themselves to eat any vegetables. One said she only felt safe when eating reheated frozen Yorkshire puddings, the main thing her mother, an alcoholic, cooked for her. Even now, the sight of vegetables nauseated her. This woman wasn't stupid. She had not failed to comprehend that vegetables are healthy. She got it; but the roots of her behaviour lay elsewhere, deep in the past.

Apart from the health implications of eating such a limited diet, it is socially awkward. Any meal in an unfamiliar setting is fraught with potential embarrassment. I spoke to another neophobic woman who said that whenever friends suggested a meal out, she had to call ahead to the restaurant to confirm that they could cook her a plain hamburger with absolutely no condiments. She ate no vegetables, though she was training herself slowly to like some fruits. When I asked why she disliked vegetables so much she laughed ruefully and said, 'I think when I was about three, my mum got fed up with me being so fussy, so she decided to let me just have the things I liked.' Which meant processed meats, chips and not much else.

The belief that tastes are a facet of personality – or genes – has dangerous consequences. If you think that children are born with certain inbuilt likes and dislikes – as fixed as eye colour – you may make no attempt to change them, because what's the point? In a 2013 journal article called 'Why Don't They Like That? And Can I Do Anything about It?' nutritionists interviewed sixty Australian parents about their children's likes and dislikes.[53] They found that parents of children who had unhealthy eating habits were much more likely to think there was little parents could do to influence their offspring's tastes, because children were just born to be difficult eaters or not.

The parents of healthy eaters made very different comments. They talked about how a child's tastes were not 'set in stone'. One of the mothers said it was possible to 'educate' the taste buds of children by exposing them to lots of different foods. Compared to the parents of unhealthy or neophobic eaters, parents of healthy eaters had a much stronger belief in their own power to influence a child's likes and dislikes. Because they believed their actions had an impact on the children, these parents did their best to create a food environment where the children could develop enough healthy likes for a 'balanced diet'. Conversely, the parents of the unhealthy eaters thought there was nothing they could do; and so, from the sound of things, they had more or less given up.

You could, of course, read this study in a different way. Not all children are equally easy to feed and there is undoubtedly a temperamental (and genetic) aspect to neophobia. Some toddlers are very much more reluctant to attempt new foods than others, no matter what parenting they receive. Maybe the parents of the healthy eaters chose to attribute their child's good habits to their influence when really it was just luck (or genes). It's easy to believe there is no such thing as genetic fussiness when your children eat well. When you are trapped in daily battles with a finicky toddler, enduring porridge thrown in the face and cauliflower on the floor, it can be irksome to listen to the smug parents whose children will 'try anything – celeriac's her favourite!'. Maybe the neophobic children really were harder to influence than the non-neophobic healthy eaters.

Nevertheless, there is strong evidence that the parents of the healthy eaters were right. Even if some of us take longer to warm up to vegetables than others, likes and dislikes are not predetermined. In most cases, it is perfectly possible not just to persuade children to eat vegetables – but to love them.

Dr Lucy Cooke spends her days trying to figure out how children's dislike of vegetables can be reversed. Cooke's research[54] – in collaboration with colleagues at University College London, notably Jane Wardle – makes her hopeful that our genetic inheritance for food preferences can be overcome. After all, she herself was once a child who didn't like vegetables, and now she is a slim, confident person who positively enjoys healthy eating, although she tells me one day at a pavement café over toasted teacakes and mint tea that she does sometimes feel deprived to think of all the foods she could eat and doesn't. 'But one mustn't say that!'

In Cooke's view, the enterprise of weaning children onto solid food should be managed with a view to setting them up with healthy likes for life. When children actually enjoy vegetables – plus a range of whole foods from all the other nutrient groups – half the battles over dinner disappear. Most parents see the aim of feeding as getting as much wholesome food into a child as possible. We focus too much on short-term quantity – kidding ourselves that if they are pacified with enough baby rice they'll sleep better – and not enough on building long-term tastes. 'The only mums we see who talk about developing a child's palate are French,' in Cooke's experience.

From four to seven months, it seems that there is a window when humans are extraordinarily receptive to flavour, but by following current guidelines on exclusive breast-feeding, parents tend to miss it.[55] Several studies have shown that when vegetables are introduced at this age, babies are more open-minded. It takes fewer exposures to persuade them to like a new flavour and the effects are long-lasting. When seven-month-old babies in Germany were exposed to a vegetable purée that they particularly disliked – such as spinach or green bean – it took only seven attempts for them to like it as much as their once-preferred carrot purée.[56] Two months later, all but 10 per cent of the children still enjoyed the once-hated vegetable, even though they had now reached an age of greater wariness. The flavour window is only fully open for a short time and seems to decline even from four to six months. A 2014 study found that when babies were introduced to a single vegetable at six months – pea purée – they ate significantly less of it than babies who were introduced to a range of purées at four months.[57]

For this reason, Cooke disagrees with the 2001 directive from the World Health Organization (WHO) that said babies should be offered an exclusive diet of breast milk for six months, with no additional solid food. This WHO report forms the basis of official guidelines to mothers in most countries, even though the statistics it was based on were mostly from the developing world, where the risks of moving away from exclusive breastfeeding before six months – such as an increased chance of gastroenteritis and faltering growth – outweigh the benefits. In rich countries, however, the norm is for most mothers to stop breastfeeding, exclusive or otherwise, well before six months. In the UK just 1 per cent of mothers are still exclusively breastfeeding at six months after birth.[58] In the US it is 18.8 per cent.[59] The main effect of the official guidelines is to hold back many formula-fed babies from experiencing any flavour except for milk from four to six months. Here, the real risk is in producing children with limited tastes who will be set up for a lifetime of unhealthy eating. As so often, we fail to see the long-term picture.

It's not that a four-month-old baby is likely to grow any better in the short term when their diet includes a spoonful here and there of veg. It's that waiting until six months to wean is to miss two months in which a child could be tasting different vegetables every day, preparing them for a recognition – and hence liking – of those same vegetables at a later stage.

The second mistake parents make – and they are encouraged to do this by those baby feeding guides with their brightly organized charts of first foods – is starting children off with bland, honeyed tastes, such as carrot, butternut squash and sweet potato. Start with those vegetables that are naturally sweet, urges the UK's bestselling author on baby food and save stronger flavours for later.[60] The really useful thing, however, would be to get a baby used to more bitter or challenging vegetables: cauliflower, courgette, spinach, broccoli, even sprouts. Weaning guides often advise sticking to a single vegetable for a whole week before switching – for fear of food allergies – but Cooke advocates lots of variety and daily changes to maximize exposure before the child enters the age of neophobia. When novel vegetables are offered to a six-month-old, the baby will often make the most dramatic expressions of horror and woe, screwing up the mouth and nose in ways that, on an adult face, would suggest torture. The hardest thing for parents is to press on with offering the food. 'We have to persuade mums to ignore the face,' explains Cooke. Unlike Clara Davis, who wanted to see how babies would eat away from parental influence, Cooke has devised pragmatic experiments recognizing that parents are part of the feeding process. Her starting point is always looking at what parents already do when feeding children and trying to find evidence-based techniques to help them do it better.

What if you have missed the 'flavour window' and are now attempting to feed a toddler who fears anything green? Is all hope lost? Cooke and colleagues found that even with school-age children, there was a great deal that could be done to change apparently fixed hatreds of certain foods. Their first revelation was that much of what manifests itself as fussiness is a response to the stressful situation of mealtimes. It can get to the extreme point where lunch itself is a 'dislike' – the pressure, the heightened emotions – no matter what is served. Cooke found that if parents could do tasting sessions with children outside meals, it could take emotion out of it. In addition, they only asked children to try pea-sized amounts of food, which reduces the feeling of pressure. 'The demands on the child have to be very low.' A whole plate of cauliflower is a horrible prospect if you don't like it. A minuscule fragment might just be OK.

Cooke helped devise a new system for encouraging more vegetable 'likes' called Tiny Tastes.[61] It was trialled in both schools and in homes and has proved remarkably effective in making children actually like raw vegetables such as carrot, celery, tomato, red pepper and cucumber. I used the scheme on my own youngest child – then aged four – and was startled how quickly it turned him from someone who said 'yuck' when he heard the word cabbage to a happy nibbler of raw green leaves. It works like this. The parent and child together select a vegetable that the child currently moderately dislikes (as opposed to feeling deeply revolted by). Each day for ten to fourteen days, not at dinnertime, you offer them a pea-sized amount. If they taste it – licking counts, it doesn't have to be swallowed – they get a tick in a box and a sticker. If not, it's no big deal; there is always tomorrow.

The usefulness of Tiny Tastes is that it provides a non-stressful way to enact the multiple exposures that we seem to need to develop new tastes. In our house, it changed the whole conversation around mealtimes, from one of stress and anxiety to something – mostly – more positive and mellow. Because he chose the vegetable himself, my child seemed to feel less trapped. Plus, he really likes stickers. Lucy Cooke said that before they started using stickers in their experiments, there would always be a few children who would refuse to take part; with stickers, participation went up to 100 per cent. Cooke's research overturns the previous orthodoxy that offering rewards for eating would make children like the food even less. Her hunch is that rewards only work first when they are not themselves food and second when the child feels they have genuinely worked for them. If you reward someone for eating a healthy food that they already like, it confuses them. But it takes a real effort for a child who dislikes raw red pepper to put that first morsel in their mouth, hence they feel they deserve the sticker.

This approach to creating new, better likes sounds almost too good – too simple – to be true. For one thing, it only addresses vegetables, which is a good place to start, but there's a lot more to a healthy diet than just greens. For many children, it is the protein foods – eggs, meat, fish – that are the hardest to love. Tiny Tastes also presumes that a child will willingly cooperate, once stickers are proposed. What about the hardcore food refuseniks? Some people have very definite dislikes with their roots in complex conditions, which surely can't be wished away with a sticker.

When children have learning difficulties or other disabilities, one of the many daily tasks they often tussle with is eating. Children who are slow to speak also tend to be slow to master the skills of eating, because there is a strong relationship between the muscle control needed for language and for chewing and swallowing. Eating can also become a problematic business for those whose condition involves rigid behaviours and routines. Those on the autistic spectrum are far more likely to have a wide range of problems with food than other children. It's been estimated that 75 per cent of children diagnosed with autism have severe feeding difficulties.[62] They may demand exclusively 'yellow' food (crisps, corn, biscuits, popcorn, fried chicken), or refuse to eat a meal unless none of the components are touching. Above all, autistic children are likely to have a very narrow range of foods that they find acceptable.[63]

Jim,* aged three and a half, was an autistic boy with serious eating problems by the time he arrived at the Penn State Hershey Medical Center in Pennsylvania. He ate only two foods, toasted cheese sandwiches and hotdogs,

---

* Names were changed to protect privacy.

supplemented with frequent glasses of milk. In addition, Jim tended to be disruptive at mealtimes, tantrumming, crying, acting out, refusing food from even his limited repertoire.[64]

But Jim was doing well compared to Kim, a five-year-old autistic girl referred to the same clinic. For a while, Kim too had eaten a limited diet of hotdogs, peanut butter, bacon, chocolate, eggs and toast. She too would cry, tantrum and throw food at mealtimes. After an illness, though, she stopped eating altogether and for six months had been completely dependent on feeding through a gastrostomy tube.

Most parents would feel overwhelmed at the thought of feeding these children and somehow broadening their horizons. I know I would. Food refusal is demoralizing at the best of times; all the more so when you are dealing with the other challenges of caring for an autistic child. If a child disliked most foods to the point that they provoked tears and rage, it would be very hard to bring yourself to do anything other than sigh and make another toasted cheese sandwich.

Jim and Kim sound like two hopeless cases. But they weren't. Within two weeks of intensive treatment at the centre, Jim's repertoire of foods had increased from three to sixty-five. Kim, meanwhile, would now eat forty-nine different foods and no longer needed the feeding tube. This huge increase in 'likes' (and decrease in 'dislikes') was achieved not through any magic but simply through a more systematic and intensive version of Lucy Cooke's Tiny Tastes system.

Therapists at the clinic engaged the children in many repeated taste sessions to expose them to pea-sized amounts of novel foods in the course of the day. Unlike with Tiny Tastes, the therapists added in an 'escape prevention' element: the child was told 'when you take your bite, you can go and play' and were not allowed to leave the room until the bite was taken. If they screamed or cried, this was ignored, but if they ate the pea-sized bite, they were praised. There were also 'probe meals' at which larger quantities of the new foods were offered – three tablespoons of three different foods – with a ten-minute time limit and no requirement to eat the food.

The results of this experiment are astonishing. To go from being fed by a tube to being able to eat forty-nine different foods is life-changing, for the whole family. A three-month follow-up showed that Jim and Kim had not lost the majority of their new likes at home. They had not slipped into the old unhappy mealtimes of before. Food was no longer a trauma to them. Both sets of parents were continuing to offer the children taste sessions outside mealtimes. Jim's range of foods was now fifty-three. This large repertoire of foods was all the more impressive considering that Jim's parents had decided to become vegetarian since the start of the intervention, the sort of change that autistic children often find unsettling. Kim's range of foods was still forty-seven. In place of a tube, she was now enjoying a wide range of different flavours and textures, without tears or rage. Autism goes along with restricted social interaction. Yet Kim's new likes placed her back in the social world of the family dinner table.

Similar work is being done by therapists at specialist feeding clinics across the world, although Keith Williams, head of the feeding clinic at the Penn State Hershey Medical Center, says that these methods are by no means standard practice. Too many feeding therapists still treat limited eaters such as Jim and Kim by offering them whole platefuls of disliked food and hoping they will suddenly decide to eat it. But when these interventions succeed, they show what huge potential there is for changing our likes and dislikes for the better through a change to our eating environment. No one is doomed to like nothing but cheese sandwiches and hot-dogs. If it's possible to train a severely autistic three-year-old to love fifty-three different nutritious foods, there's hope for us all.

The trouble is, though, that most of our food environment influences us in an opposite direction. Every day, children are exposed to messages – whether on giant hoardings and TV ads or from looking in friends' lunchboxes – telling them that they should like the very foods that will do them the most harm.

Karl Duncker's 1930s experiments on children's likes and dislikes are much less well known than Clara Davis and her feeding orphanage. But they offer just as great an insight into how our tastes are formed, almost in spite of ourselves, by forces we are only dimly aware of. While Davis was interested in what tastes look like stripped of the normal social influences, Duncker wanted to pinpoint how those influences actually work.

In 1936 Duncker (born in Leipzig in 1903) was a promising young Gestalt psychologist exiled from Nazi Germany – where his parents were prominent communists – to Britain where he continued his work. One of his great philosophical interests was pleasure and what causes it. His definition of the pleasure of anticipation was a child who 'has been told that he is soon to have a piece of candy . . . glowing all over with happiness'. In one of his papers, Duncker asked why eating a fine juicy beefsteak could cause such delight; he decided that it wasn't just that it took away the pain of hunger. It was the sensory enjoyment of biting into it, and the feeling it gave that 'life is grand'.[65]

On arrival in Britain, Duncker set himself the task of investigating the role of social suggestion in forming food preferences. Given that likes and dislikes varied to an 'astounding degree' among different cultures, he realized that there must be a process of social influence at work. His mission was to unravel the psychological processes by which likes were formed.

Duncker's experiments involved children from Somers Town nursery school in London NW1, which was then a poor district of London.[66] The first experiment was a simple one. Boys and girls aged between two and five were asked to make a food selection from carrots, bananas, nuts, apples, bread and grapes. What Duncker found was that children were far more likely to select the same foods as one another if they made their choice in the presence of other children, than if they were alone. For children younger than twenty-seven months there was a wonderful 'social indifference': 'when they had fixed their minds upon the food, nothing else seemed to exist.' Above that age, however, there was a marked tendency to copy the likes of other children, especially if the child who selected first was just a little older. There was one pair of girls where one was an extroverted five and the other a shy four. Before choosing her food, 'B would always send some furtive glances over to A as if for reassurance'.

We've all seen this kind of peer influence at work. If you offer a snack to a group of young girls, they will often tie themselves in knots second-guessing what the others will go for before making up their own mind. You don't want to be a lone wolf eating popcorn when everyone else has opted for toast. Duncker's findings about social suggestion when eating have since been confirmed by at least sixty-nine separate experiments.[67] This is a very robust phenomenon. Depending on the influence of those who share our meals, we may eat faster or slower; we choose different foods; we manage larger or smaller portions.

Duncker's second experiment was more dramatic. He took two substances. One was a white chocolate powder flavoured with lemon – a very luxurious commodity in 1930s Britain and 'decidedly pleasant'. The other was valerian sugar coloured brown, valerian being a herbal root traditionally used as a sedative: a very bitter and medicinal flavour that Duncker called 'rather unpleasant'. He then asked the nursery teacher to read the children a story about a hero, Micky, a little field mouse, who hates one food – 'hemlock' – and loves another – 'maple sugar'. When Micky discovers maple sugar in a tree, he realizes he has never 'tasted such good stuff before'. But the hemlock bark is 'sour and disgusting'.

After the story, the children were then asked to taste some actual 'hemlock' – which was really the delicious white chocolate powder; and 'maple sugar' – which was really the unpleasant valerian sugar. The deception did not exactly work. Many of the children recognized that the 'hemlock' was really chocolate. Yet when asked to choose which substance they preferred, 67 per cent of them opted for the nasty-tasting 'maple sugar' because of the positive associations in the story (only 13 per cent chose it in a control group with no story).

Can our likes and dislikes really be so easily influenced? Apparently so. Duncker's experiment shows that a simple story is enough to make children forget – for a time – that they like chocolate. For Duncker himself, having witnessed Hitler's rise to power, it was no surprise that human beings are suggestible in their 'likes' or that social forces can make them suppress their natural impulses. At the time Karl

Duncker was doing his peaceful experiments with children and chocolate, his younger brother Wolfgang was living a precarious life in exile in Moscow; he was arrested during the Great Purges of 1938 and died in the Gulag. Duncker himself had lost his academic position in Berlin in 1935 for having once been married to a Jewish woman.[68] 'If educated adults,' wrote Karl Duncker, 'can be made to discard their ingrained preferences because the leader has contrary ones, why should children prove [harder to influence] – even in such a vital domain as food?'[69]

Given his background, Duncker had a strong sense of how those with power manipulate the powerless. To him, a child being manipulated to change their ingrained food likes was in a similar position to the population of Nazi Germany.

Duncker's findings are deeply worrying. If just one story about a not very inspiring mouse hero could make children change their likes to such a degree, what are the effects of a daily barrage of advertising stories, in which godlike athletes are shown drinking sugary beverages and the least nutritious cereals are those with the cutest characters on the box? 'Don't trust that tiger! He's a bad tiger!' I used to tell my son as we walked down the cereal aisle.

What can any one of us do in the face of such social pressure? Duncker offered himself up as an example of how individuals could train themselves to new likes, despite their social prejudices and circumstances, through a kind of 'inner reorganisation'. When he arrived in Cambridge from Germany, Duncker was appalled by the prevalence of something called 'salad cream': a sharp condiment beloved in the British Isles that has the texture of mayonnaise but the acrid taste of spirit vinegar. Like many mass-market foods, it has a devoted following among those reared on it, but to Duncker, who wasn't prepared for the taste, salad cream came as quite a shock.

> Suffice it to tell just one personal experience. When I first came to England, I was made to understand that raw green salad leaves could be made into 'salad' with the aid of a bottled substance of yellowish color, called salad dressing. It looked like mayonnaise; I expected mayonnaise – and I dare say I was deeply disappointed. No, I did not like it. But as I did not like raw leaves either, I was therefore prompted to adopt the most favorable and adventurous attitude. I tried again, and I still remember the day when suddenly I discovered that this was not an unpleasant variant of mayonnaise but a kind of mustard which was not unpleasant at all. Thus by accentuating the mustard potentiality and suppressing the non-mayonnaise aspect, I came to like it.[70]

Like Lucy Cooke, Duncker knew that there is huge scope for changing our likes and dislikes: not all of them, for sure, but enough to make the difference between a good diet and a bad one. Whether you are a PROP taster or not; autistic or not; neophobic or not; fussy or not; a foreigner or not; genes are never the final reason why you like the particular range of foods you do. When a boy likes nothing but cornflakes, it says less about him than it does about the world he lives in.

It would help if we stopped seeing our personal likes as such a deep and meaningful part of our essence. There are many things about ourselves we cannot change, but the majority of food likes do not fall into this category. Our tastes are learned in the context of immense social influences, whether from our family, our friends, or the cheery font on a bottle of soda. Yet it's still possible, as Duncker showed, to carve out new tastes for ourselves. We can put the impressionable nature of our likes to good use. If we expose ourselves enough times to enough different foods, we may find, like Duncker, that the flavours we once disliked have now miraculously become likeable.

In Duncker's case, sadly, taking what he called a 'favourable and adventurous' attitude to food was easier than taking a favourable attitude to life. By the time he was doing his experiment with children and white chocolate, he had been suffering from deteriorating mental health for the best part of a decade. Duncker missed life in Berlin, but knew he could never return while the Nazis were in power. Unlike his tastes in salad, this situation was intractable. In 1938 he emigrated from Britain to the United States to take up a job at Swarthmore College. It was there he committed suicide in 1940, at the age of thirty-seven.[71]

BEETROOT

Every culture seems to have certain challenging vegetables that children find hard to love at first bite. And at second. And third. In Brazil, it is okra (the sliminess). In France, it may be turnips (the bitterness). In lots of countries, it is beetroot (the purpleness).

There are plenty of reasons to find beetroot off-putting. There's the curious taste, reminiscent of earth and blood (the culprit is a chemical compound called geosmin). Also, the texture, which in its cooked form is neither crunchy nor soft but fleshy. Most of all, there is the shocking colour that bleeds inescapably over everything on your plate.

Yet among sophisticated adult eaters, beetroot is often a special favourite. It thus offers a case study in how we can learn new tastes. It's not just that people learn to tolerate beetroot: they switch from dislike to adoration. Since the 1990s beetroot has been a beloved item on restaurant menus, often paired with goat's cheese. Adult beet lovers enjoy the very qualities that children find so awful: the earthy taste and meaty texture and, most of all, the bright crimson pigment, which can dye a whole pan of risotto a joyous pink.

Between the beet haters and the lovers, there is a gulf. Some of it can be explained – as with many other dislikes – by the form in which we first encounter the outlandish purple vegetable. Childhood memories of vinegary pickled beetroot do not help. When someone learns to love beetroot, it is often because they have been given a taste of these roots in a new and more appealing form, when eating out: a fresh and vibrant beet and orange salad, say, or a moreish deep-fried beetroot crisp.

Regardless of cooking method, however, there does seem to be something in strong vegetable flavours such as beetroot that people take longer to hit it off with. In one study, seven- and eight-year-old children from the Netherlands were given tastes of pure beetroot juice every day for fourteen days, the kind of 'exposure' that in theory should lead to liking. But at the end of the fortnight, they continued to find the beetroot taste 'too intense'.[72]

Maybe it is the sense of achievement at having conquered an aversion that makes adult beetroot fans flaunt their enjoyment so overtly. Foodies trumpet their love of the hated vegetables of childhood: cauliflower and Brussels sprouts join beetroot as dinner party favourites. But beetroot eaters are not just showing off. It is possible to reach the point where these complex, bitter flavours deliver more pleasure than the simple blandness of mashed potato.

The psychologist E.P. Köster has shown that one of the beneficial effects when children are exposed through 'sensory education' to a wider range of flavours is that they start to love complexity and be bored by simplicity.[73] Given time and enough attempts, we actively seek out those foods – like beetroot – whose charm is not at first obvious.

CHAPTER 2:

Memory

The women have a lot to talk about;
they remember their homes,
and dinners they made.

Poem written by EVA SCHULZOVA, aged twelve,
in the Terezin Concentration Camp

When Abi Millard was four, her mother Dawn started to notice that Abi was acting strangely at mealtimes. She seldom seemed hungry and often put down her fork after a bite or two. Though generally happy and well behaved, Abi was, in Dawn's words, 'a nightmare' when the family went out to eat with friends, 'messing around and not eating her dinner'. They took her to the doctor who diagnosed congenital anosmia: an inability to smell, which also means an inability to taste food properly, given almost all of what we call 'taste' is really flavour perception through the nose.

To consider anosmia is to see how central food memories are, both to the way we learn to eat and the way we relate to the world. When I met her, Abi Millard was nine. In most respects she is a self-assured, happy girl. She likes swimming and Tae Kwon-Do. She lives with her mum and dad in a rural village and goes to the local primary school. Yet her experience of life is different from most. Without the ability to smell or perceive flavour, Abi experiences food more or less as pure texture. Blindfolded, she can't tell the difference between houmous and strawberry yoghurt. Salad leaves tickle her throat and tomatoes are slimy, though she will eat broccoli and carrots and peas. She has few of the drives that motivate most people to seek out certain beloved foods because she lacks the memories that would make her expect them to be rewarding. Dawn says she lacks any real enjoyment in food – except for one time when they were out at a restaurant and Abi ate a gammon steak and said, 'That's lovely', perhaps because it was intensely salty (Abi can detect strong concentrations of salt or sugar on her tongue, but without any of the flavour nuances). Dawn worries that when Abi is grown up, she may forget to eat. The illness is also isolating: when Abi's friends at school talk about favourite meals, it's hard for her to join in. She has no idea what a batch of warm vanilla shortbread smells like; or chocolate; or garlic. She has no memory of the taste of her own mother's cooking.

It is extremely rare to have anosmia from birth, as Abi does. What is far more common is to develop anosmia later in life, often after a head injury (though it may also be caused by sinusitis, nasal polyps, dementia, chemotherapy, stroke, liver disease or sometimes for no clear reason at all). At a conference organized by the anosmia support group Fifth Sense in spring 2014, anosmia sufferers spoke of how doctors were often dismissive of their condition: 'be thankful you're not deaf' was a common refrain. But it's hard to be thankful for a malady that separates you from the food memories that define so much of who you believe yourself to be. One woman at the conference who had been in a cycling accident talked about how her marriage had broken down after she became anosmic. Her relationship could not survive her inability to share her husband's continuing pleasure in food. Before the accident, they had both loved throwing dinner parties, but her husband couldn't understand that elaborate cooking now did nothing for her. Every meal was a cruel reminder of what she had lost. The predicament of those who are born with anosmia, like Abi Millard, is that they can't share the pleasurable food memories that the rest of us have. The predicament of those who develop anosmia later in life is that they have the memories, but no means to access them. They are cut off from their own past.

It was a bright spring day in San Francisco in 2011 when Marlena Spieler, a food writer with more than twenty cookbooks to her name, was knocked down by a car at a crossing. Both her arms were broken and she suffered concussion. As the initial excruciating pain gradually lessened, Marlena – a sunny optimist with Marilyn-platinum hair – noticed another injury, which to her was far worse than the broken limbs. The head trauma had damaged the nerve connecting to her olfactory bulb – the part of the brain that interprets flavour – and she could no longer enjoy food. Coffee had been one of Marlena's great pleasures since she was very young. Now it was tasteless. 'Cinnamon drops, a childhood favourite, were bitter, horrible,' she wrote in the *New York Times*. 'Tamales were as bland as porridge. Bananas tasted like parsnips and smelled like nail polish remover.' As for chocolate, it was 'like dirt'.[1]

I got to know Marlena in 2002 on a press trip to Parma organized by the consortium of producers of Prosciutto di

Parma. For three days we ate ribbons of pink salty ham at every meal and Marlena, with great Californian ebullience, talked about the foods she loved the most. These were long – very long – conversations. She spoke of artichokes and lemons; earthy dried mint and pungent truffles; bread and cheese; and how she would rather be in Italy than anywhere in the world. She ate with a dainty slowness, as if trying to extract the essence from each bite.

After the accident, she could still perceive the heat of pungent spices such as mustard, Aleppo pepper or cinnamon, because her trigeminal nerve, the part of the body that tingles when you eat hot food, was not damaged. But without the flavour to offset it, the tingling sensation was offensive. Her beloved cinnamon felt abrasive. Meanwhile, she developed new cravings, for intensely sweet desserts and for fish. When I first knew Marlena, she was indifferent to desserts and hostile to fish. Now, suddenly, she had yearnings for smoked mackerel and anchovies. She also developed a very sweet tooth. A scientist working on flavour and the brain told her that this might be because she could no longer detect the qualities in fish and sweets that she had previously been disgusted by. They only seemed desirable because she couldn't recognize the ways they had once repelled her.

People sometimes speak of anosmia as 'loss of taste' but damage to the taste buds themselves is actually very rare. More than 90 per cent of cases of taste-related disorders involve a weakening or loss of the sense of smell. The taste buds in our mouths only supply a fraction of the complex pleasures that we enjoy as 'flavour'. The rest is perceived via our noses, through something called retronasal olfaction. We smell coffee by breathing in – is any scent better than a warm bag of freshly ground beans? But we taste a cup of coffee by smelling it *backwards*, or retronasally. The hundreds of chemical compounds that go together to make up the flavour of a particular blend and roast of coffee travel to the back of our mouths and sneak backwards through the nasopharyngeal passage into the nasal cavity.[2] As we sip and swallow, we are not conscious that the splendid flavours – the nuttiness of the roast, the notes of cherry and peach – are created in the nose, not the mouth. This spectrum of retronasal joy is lost to anosmia sufferers. All they have left are the harsh and basic tongue-notes of sweet, sour, bitter and salty. Like Abi Millard enjoying a salty gammon steak, anosmia sufferers often seek out extremely salty or sweet foods to compensate for the loss of flavour.

Anosmia is a surprisingly common problem – as many as 2 million people in the US have some form of smell or taste disorder.[3] It is not a trivial disability. The ability to pick out the jasmine aroma in a cup of espresso or to spot the difference between grapefruit and pomelo might seem of little importance to anyone except for food writers. But medicine and neuroscience are now starting to recognize that anosmia can be an extremely traumatic condition, and not just because of the danger in emergencies of not detecting the smell of smoke or gas. Sufferers often end up depressed and malnourished. Without flavour, the motivation to eat is lost. When nothing can be smelled, there is a yearning for familiar tastes that can never be satisfied. Christmas goes by without the background aroma of turkey or spice; summers are no longer marked by the perfume of strawberries and cut grass. Sufferers often describe it as a deep loss. Duncan Boak, the founder of Fifth Sense, who became anosmic after a head injury, said that he feels as if he is looking at life through a pane of glass.[4]

Part of what is missing for the anosmia sufferer is the safe place of childhood, which the rest of us can return to whenever we eat the foods we have always loved. A couple of years on from the accident, Marlena Spieler found that glimmers of her former responsiveness to flavour were returning. Depending on the level of damage to the brain, some anosmia sufferers do recover. Marlena slowly trained herself to love chocolate again, starting with the blandest milk chocolate and working onto 70 per cent cocoa solid dark chocolate. Occasionally, her morning coffee gave her pleasure rather than just blankness. We met for lunch at an Italian restaurant and she seemed well, exclaiming over the blood orange slices in our cocktails and nibbling on a deep-fried sage leaf. But even as her flavour perception was improving, she continued to feel unsettled, she said. It wasn't just that food tasted bad. It was that, she told me, she no longer felt quite 'like Marlena'. As she explained to the BBC Radio 4 *Food Programme*, our sense of taste is something that anchors us to the person we have always known ourselves to be: 'Your world has a certain taste. Your mother makes something a certain way. You're used to certain flavours in your life and if you take that away, you start saying who am I?'[5]

Memory is the single most powerful driving force in how we learn to eat; it shapes all our yearnings. Sometimes the memories are very short-term ones – for instance, whether or not we've just eaten. In one study, when a profoundly amnesiac patient was offered another meal only minutes after he had completed the first, he willingly took it.[6] Minutes after that one, he ate a third meal. Only when a fourth meal was offered did he refuse, telling the experimenters that his 'stomach was a little tight'.[7] This suggests that having a conscious memory of our last meal matters as much as hunger in determining how much we eat.

For most of us, though, the food memories that really matter go much further back. You may not be able to remember what you had for lunch last Tuesday, but I bet you can recall the habitual meals of childhood; the breakfast you were given for a weekend treat and the way bread tasted in your house. These are the memories that still have emotional force years or even decades later.

Such memories, conscious or unconscious, are what drive us to seek out the old habitual foods – particularly packaged foods – even if, judged objectively, they do not really taste nice or do our bodies any good. There have been experiments done with rats and mice where the animals are given dopamine blockers, drugs that interfere with the part of the brain that governs reward. These drugs take away much of the chemical reward of eating food. Yet the dopamine blockers do not extinguish the rodents' food-seeking behaviour, at least not straight away. At first, the animals continue to press the lever (or run through the alleyway, or whatever the task might be) and eat the pellets, even though the dopamine blocker means that the food no longer offers the same gratification.[8] Next, they carry on pressing the lever to earn the pellets, but do not eat them. Finally they stop pressing the lever, indicating that at last their desire for the pellets has gone. The interesting thing is that it takes so long for the desire to fade. As the neuroscientist Roy A. Wise has observed, it is only when 'the memory of the reward is degraded through experience that the desire is lost'.[9] The craving for the pellets is more a function of memory than of how they taste. Memory propels human food urges in much the same way. As we traverse the supermarket aisles in a trance-like state, we are like rats in an alleyway, steered to this or that food by memories of rewards long gone.

One of the reasons that we do not usually think of our tastes as learned is that most of the learning tends to happen in the very early years of life; and then it stops. For those of us who believe in personal development, it is depressing to learn that a person's food 'likes' aged two generally predict their tastes at twenty. In 2005 researchers in Turkey interviewed nearly 700 undergraduate students and their mothers.[10] The mothers were asked about their children's eating habits when they were two and the students were asked about how they ate now. There was a remarkable continuity between then and now. The students who were 'picky eaters' as children still described themselves as picky eaters. The ones whose mothers recalled that they always ate too much still did so. And the three people in the study who 'never' ate vegetables as children still had no vegetables in their diet. So much for putting aside childish things.

When we talk of memory and food, we generally assume that nostalgia is a phenomenon that occurs late in life – like Proust being transported to his youth by a madeleine dipped in lime-blossom tea. But food memory is there from the start. Even babies have nostalgia! It's a large part of how we learn to eat. The foods parents give to babies provide them with powerful memories that trigger lasting responses to certain flavours. This process begins before birth. We are all born with echoes of our mother's diet, which mean that no one is a totally blank slate when it comes to flavour. We arrive predisposed to respond to certain foods by our experiences in utero.

It's hard to know what a newborn thinks about taste, since we can't exactly ask them. Or rather, they can't exactly

answer. But in 1974 the Israeli doctor Jacob Steiner realized
that a baby's reactions to the basic tastes of sweet, sour,
salty, bitter could be gauged by their facial expressions,
which are vivid and mobile, even in the first week.[11] Steiner
took babies just a few hours old and offered them a range
of tastes on a cotton swab, filming their facial expressions.
When given salt, which you'd think might make them cry, the
babies surprisingly showed little reaction, continuing to look
expressionless (a liking for salt only emerges later, around
four months). But all the other basic mouth-tastes produced
strong reactions. The sour swab made the babies pucker their
lips. Bitterness provoked an expression of abject distress and
an open mouth, as if trying to spit or vomit it out. As for the
sweet swab, Steiner found that it produced a dreamy look
of 'relaxation' with an 'eager licking of the upper lip' and
even a 'slight smile' – and this at an age when babies are
not supposed to be capable of smiling. Such is the power
of sugar.

The test has since been repeated many times, with similar
results. What it confirms is that, as we have seen, all human
babies, from Sweden to China, have a strong innate preference for sweetness and a dislike of bitterness and sourness.
Basic tastes are not a question of memory: we are hard-wired
to think sweetness is wonderful and that bitterness is scary.
No one has to learn these simple tongue-reactions. But flavour is another matter. Flavours – these memories generated
backwards through our nose – are *all* learned. What we think
about flavour in all its myriad forms, from toasted cumin to
sea bass, from parsley to spaghetti carbonara, is not fixed.
Each of us will have a different bank of memories and feelings about these; and it exists from day one, if not before.

Taste buds appear at seven or eight weeks of gestation.

Already, by thirteen to fifteen weeks, the taste buds are mature. A thirteen-week-old foetus weighs maybe an ounce, with no fat under the skin, no air in the lungs. Yet already they can not only swallow but *taste*, and these sips of fluid leave memories.

In 2000 some French scientists did a remarkable experiment showing that newborns arrive in the world with a memory of how their particular amniotic fluid tasted.[12] The mothers studied came from the Alsace region where strong-tasting anise sweets are a local delicacy. Some of the women had eaten anise regularly during pregnancy and some had not. The babies were tested straight after birth and four days later, having tasted nothing outside the womb but milk. When an anise odour was wafted in front of them, the babies born to anise eaters showed a marked and 'stable' preference for anise. They turned their heads towards the anise smell, sticking their tongues out with a licking gesture. They remembered it and apparently it pleased them.

Further experiments have confirmed that other strong flavours such as garlic can also find their way into amniotic fluid. In one study, women agreed to swallow garlic capsules forty-five minutes before they were due for an amniocentesis; when it was tested, their amniotic fluid smelled garlicky.[13] Babies born to voracious garlic eaters will have been floating in a sac of garlic water for nine months. It has been shown that babies exposed to garlic before birth are more likely to enjoy garlic in food later on. Likewise, mice whose mothers had been fed on artificial sweeteners when pregnant had an exaggerated taste for sweetness.[14] Pregnant rats fed on junk food – including savoury snacks, sweetened cereals and chocolate-hazelnut spread – had babies who also selected these foods over regular rodent pellets, though the babies' preference for junk was lessened if the mothers switched to a healthier diet during lactation.[15]

The flavours our mothers ingest most regularly can become like mother's milk to us. Julie Mennella and Gary Beauchamp are biopsychologists working at the Monell Chemical Senses Center in Philadelphia who have done a series of experiments on how flavour in utero and in breast milk leaves children with lasting memories and preferences for certain foods.[16] One of their most celebrated studies, from 2001, involved carrot juice. The babies of a group of mothers who drank carrot juice during the last trimester of pregnancy and again during the first two months of breast-feeding were predisposed to like the flavour of carrot. When the babies were weaned onto solid food, several months after the mothers stopped drinking the carrot juice, they showed a marked preference for cereal flavoured with carrot juice over plain cereal flavoured with water.

The early exposure of babies to flavour – both in utero and through milk – works as a kind of 'imprinting', as Gary Beauchamp puts it.[17] We become emotionally attached to these early aromas. As we saw in Chapter One with the 'flavour window', younger babies are more open than older ones to new tastes. When it comes to weaning, this is an argument for ignoring the advice on exclusive breastfeeding for six months and offering early, varied bites of vegetable purées between four and six months. When it comes to the pre-food stage, however, flavour may be one of the strongest arguments for mothers attempting to breastfeed, at least for the first few months, to eat as varied a diet as possible while doing so. Some psychologists suggest that instead of saying to mothers: 'breastfeed for the baby's good' healthcare advisers should say: 'breastfeed for your own good' because you are likely to have a baby who is less fussy to deal with in the early stages of eating.[18] Then again, I've known children (not my own) who've gone from formula milk at four months to black olives and spinach tart at twelve months, so it doesn't always follow.

It is curious that we talk so little about the flavour of formula, given that it is the main food many babies taste for that crucial first year. Because any given brand of formula milk does not vary, it seems to have an even greater 'imprinting' power than breast milk. Babies who cannot tolerate regular cow's milk formula are sometimes given special 'hydrolysate' formula, whose proteins have been broken down (hydrolysed) to make them more digestible. To adult tastes, these formulas taste especially nasty, with a sour cheesy tang and a strange hay-like odour. Mennella and Beauchamp followed children who had been fed with two different hydrolysate formulas.[19] Objectively, both of these milks tasted equally unpleasant. But to the infants, the particular formula they had been assigned – sour or not – taught them how food should taste. When the two brands were switched round, the infants drank less: they preferred their own bad-tasting formula to the other one. More strikingly still, children aged four to five who had been fed on these sour-tasting hydrolysates as babies showed more positive feelings about sour tastes and smells than children who had been fed on breast milk or regular formula. This is vivid proof that *anything* can start to taste good if you have enough positive memories of being fed it by a parent. The obvious implication is that formula-fed babies would benefit from having their milk flavoured with vegetables.

Formula can never match the myriad benefits of breast milk, which range from lessening the risk of eczema and ear infections to reducing the likelihood of type 2 diabetes later in life to promoting healthy gut microbiota. But in the developed world, as we've seen, most mothers are unable or unwilling to breastfeed exclusively for the first six months. With each of my own babies, I gave it up for one reason or another (illness, work, bereavement and a child with feeding difficulties) at three months. Until they were a year old, when they were old enough for regular whole cow's milk, I'd have been glad to buy formula that was mildly scented with a range of green vegetables, just enough to give them a memory of spinach when the time came for them to try veg for real.

Instead, in many countries, formula milk has been flavoured, if at all, with vanillin, the artificial vanilla flavour that goes into industrially produced sweet foods, from ice cream to biscuits to cake. Vanilla milk has a long history. Back in 1940, the head nurse of the children's hospital in Philadelphia recommended tempting reluctant feeders with three drops of vanilla essence in each bottle.[20] Internet forums suggest that there are still many desperate parents who resort to vanilla extract when a baby rejects the bottle.[21]

Since 1981, international food standards (the Codex Alimentarius of the World Health Organization) have stated that no flavourings should be added to infant formulas aimed at newborns. But vanillin is still a key ingredient in many of the 'toddler milks' marketed at children aged one and over. In China, vanillin is prohibited in infant formula but it continues to be illegally added by many manufacturers. In 2014, a team of chemical analysts found vanillin in four out of twenty samples of infant formula randomly purchased from supermarkets in the city of Wenzhou.[22]

Of all the flavours you could think of with which to

'imprint' a child, this is possibly the least useful from a health standpoint (except, perhaps, for chocolate: in 2010 the American company Mead Johnson withdrew its 'premium' chocolate-flavoured Enfagrow toddler milk amid complaints from leading nutritional scientist Marion Nestle that it was training kids to 'like candy').[23] The effects of vanilla milk are lasting. In 1999 some researchers in Germany tested the effects of the vanilla that had been in German 'bottle milk' for some years.[24] They asked 133 people to try two different ketchups, one of which was straight-up tomato ketchup and the other, bizarrely, had been flavoured with vanillin. (The reason the researchers chose ketchup was precisely because it is not normally associated with vanilla.) Of the respondents, the majority of those who had been breastfed had a preference for the pure ketchup, while the majority of those who had been reared on vanilla formula preferred the strange vanilla ketchup. Their baby milk had brainwashed these unfortunate people into thinking that vanilla made everything taste better.

Clearly, spinach milk would be a better plan, assuming it could be made safe for tiny infants. It will probably never take off, though. Over time, the odds are that babies would accept it and even prefer it, just as the hydrolysate babies with their bad-tasting formula think that milk is meant to taste sourish and cheesy. It's the parents who would find vegetable milk hard to accept. We want our babies to have milk that corresponds to our own memories of childhood. Manufacturers know that you can only sell baby food by making it appealing to adults, which is why baby rusks are sometimes sweeter than doughnuts and why for decades, until it was banned, jars of baby mush came seasoned with MSG, to give it more savoury taste. When vanilla is found in baby foods, it has been put there to attract not the children themselves – who, as we've seen, can become emotionally bonded to flavours that are strange, sour or strong given the right memories – but to please adults. The babies are not the ones who buy the food. It is the grown-ups' memories that the food companies are trying to appeal to.[25] As they warm the sterilized bottle, parents sniff their baby's milk; or maybe take a tiny sip. It is they, not the babies, who have memories of how childhood milk *ought* to taste: creamy and sweet, like milk left behind in the cereal bowl.

Do you remember your first passion fruit, your first avocado, your first Thai green curry? Such flavour memories can seem inconsequential, the stuff of gastronomes. 'Ah yes, it was in Marseille in 1987 that I first tasted an authentic bouillabaisse.'

Yet, from the perspective of neuroscience, food memories are not something slight. Registering different flavours is one of the main ways that our bodies interact with the world around us. Amazingly enough, the human olfactory bulb is the only part of the central nervous system that is directly exposed to our environment, through the nasal cavity. Our other senses – sight, sound and touch – need to travel on a complicated journey via nerves along the spinal cord up to the brain. Smell and flavour, by contrast, surge direct from plate to nose to brain.

Conventional wisdom used to be that humans have rather a weak sense of smell, compared to that of other animals: dogs, say (witness the fact that we don't have sniffer humans at airports). But recent research suggests the contrary. We may not have a bloodhound's ability to track a scent, but our olfactory discernment is second to none. We can detect a drop of Worcester sauce in a glass of tomato juice; or the scent of fear in another person's sweat.[26]

When I say that we discern smells and flavours, what I should really say is that we create them. Flavour is not actually in food, any more than redness is in a rose or yellow is in the sun. It is a fabrication of our brains and for each taste we create a mental 'flavour image', in the same way that we develop a memory bank of the faces of people we know. The difference is that whereas faces fade when you haven't seen them in a while, flavours and smells have a way of lodging themselves indelibly. What you taste as a child is still there in your adult brain, even if you haven't thought of it for years. The Norwegian Trygg Engen, the 'founding father' of the study of smell and memory, characterized our sense of smell as 'a system designed not to forget'.[27]

In 1991 the biologists Richard Axel and Linda Buck discovered that olfactory receptors – cells in the nose that detect odour molecules – make up the largest single family in the human genome. Out of around 19,000 genes, Axel and Buck found, nearly a thousand – 5 per cent – are olfactory receptors. Their research finally unlocked some of the mystery of how humans can remember and discriminate between so many flavours and smells (and, thirteen years later, won them a Nobel Prize).

What makes the human system of olfaction so sophisticated is not just the receptors themselves, but the way they interact with our large brains. Each receptor cell is fairly specialized: it can detect only a small number of substances. But when you smell or taste something – a loaf of freshly baked bread, say, or lemon zest sprinkled over a stew – the receptors send messages to the olfactory bulb in the brain. Here, each flavour becomes encoded in its own particular pattern in a part of the olfactory bulb called glomeruli. A glomerulus has been described as a 'detection point par excellence'. Each and every time you taste or smell something, the relevant glomerulus will take a snapshot of it. These snapshots show up in the brain as patterns, like a map.

Humans can distinguish around 10,000 separate smells, estimates Linda Buck. We walk into the house and instantly know that someone is cooking roast chicken for supper and that they decided to stuff it with rosemary instead of thyme. Our olfactory systems have an immense power to discriminate between different flavours. Molecules that look near-identical to a specialist chemist in a lab will be easily distinguished by an ordinary person who smells them. Our brains will also interpret the same chemical in radically different ways depending on how concentrated it is. Buck and colleagues note that a 'striking example is a substance called thioterpineol, whose odor is described as "tropical fruit" at a low concentration, as "grapefruit" at a higher concentration, and as "stench" at a still higher concentration'.[28]

Once we move beyond smell to consider flavour, however, the images processed in our brains become vastly more complex. In addition to the odour signals from our noses – this coffee is good! – there will be taste signals from the mouth – oh, but it's bitter! – as well as feelings of texture – smooth crema! – and temperature – that burned my tongue! The experience of tasting food is far more multi-sensory than is the case with hearing, sight or touch, which is why it requires the most sophisticated part of our brain to process it. In fact, eating is influenced by hearing, sight and touch as well as flavour: we prefer apples that crunch loudly, steaks that look blood-red, sauces so smooth they seem to caress the inside of our throats.

If there are 10,000 smells, the number of different flavours that our brains can potentially create is infinite. Professor Gordon M. Shepherd, a biologist based at Yale University, has coined the term 'neurogastronomy' to explain our brain's unique flavour system.[29] In Shepherd's view, complex flavour recognition is at the core of human identity, separating us from other mammals. Cats cannot even detect something as basic as sugar – they lack a taste receptor for sweetness. Humans, on the other hand, can differentiate fake maple syrup from real maple syrup; Coke from Diet Coke. Shepherd notes that the images humans build up of different flavours are processed in the prefrontal cortex, the area of the brain that is most important for decision-making and abstract thought; but also memory. Shepherd's work has shown that the human brain can potentially generate any number of flavours 'since every soluble body has a special flavor which does not wholly resemble any other'.[30]

The way our brains interpret flavours speaks to the human love of patterns. Professor Shepherd and colleagues have done experiments using fMRI and other brain scanning technologies to show that different flavours register as different patterns in the brain. It is startling to see scans of these flavour maps and realize that there is a separate place in our brains for bananas and Cheddar cheese or that strawberries and sugar show up as dots in similar locations. The way our brains map flavour is similar to the way that we perceive visual images. When we 'see' something, what we are actually doing is creating an abstract 2D representation of it, with some features enhanced and others suppressed. By the same token, when we put food in our mouths, the flavour molecules that drift to our nose are turned into abstract patterns in the brain. These patterns help us to recognize the food when we taste it again. Our olfactory receptors give different patterns to the sweet and the savoury; the rotten and the fresh. The receptors also modify the patterns depending on what is happening in the rest of the body: whether we are happy or depressed or nauseous.

Through these patterns, our brains make sense of the bewildering world of flavour. Take umami, the so-called fifth taste, which corresponds to the savoury qualities in meat, cheese and certain vegetables such as tomatoes or broccoli. Umami is what gives mushrooms their oomph and the reason it's so hard to stop pouring gravy on your potatoes. We all have neurons that are specifically tuned to umami. Yet by itself, umami – which is made in artificial form as MSG – doesn't really taste of much. It is only in conjunction with other flavours that it becomes delicious. We can see this from neuroimaging studies. When glutamates are tasted in conjunction with a savoury vegetable odour, they generate far more brain activity than when the two flavours are tasted separately. The whole is more than the sum of its parts. This makes sense. Our brains are smart enough to see that a dish of Asian greens with soy sauce warrants a more sizeable flavour image than the same greens and soy sauce eaten separately.

What is most significant about our flavour images is the way that they lead to what scientists call 'images of desire'. Once we have a memory in our heads of a flavour we love, we build up 'images of desire' as we seek to acquire it again. In 2004 researchers put subjects on a bland diet and asked them to imagine their favourite foods. Just thinking about these beloved dishes created a response signal in the hippocampus, insula and caudate – the same areas of the brain that are activated during drug craving. Canadian researchers found that people who described themselves as 'chocolate cravers' showed different brain activity when eating chocolate than self-diagnosed non-cravers. The cravers' brains continued to respond favourably to pictures of chocolate long after their bodies had reached a point of fullness. Neuroscience confirms that chocolate means more to some people than others.

To anticipate pleasure in the next meal – something that can take up the greater part of the day, in my experience – is always a form of memory. And each mouthful recalls other mouthfuls you've eaten in the past. It stands to reason, therefore, that the flavour patterns in each of our brains are highly dependent on all the things we've tasted in the past, especially during childhood. Among North Africans settled in France, fresh mint tea, often served in ornate teapots, is a way of life. Children grow up with that familiar herbal steam rising from the table as adults sit and talk. A particularly refreshing mint tea is served in the courtyard of the Mosque in Paris, a tranquil place to retreat on sweltering days in the city.

For French Algerians, mint tea is imprinted on the mind, in a way that doesn't hold true for the non-African French population. In 2009 a group of subjects, half of them 'Algerian-French' and half of them 'European-French' were asked to smell mint and say what they thought of it. All of them – French or Algerian – found it pleasant and all of them correctly identified it as mint. But when gold electrodes were attached to the scalp, the Algerians showed a significantly greater level of neural activity in response to the mint than the Europeans. Because of the mint tea they drank at home, the smell induced a different cortical pattern in the brain. Put simply, mint was a flavour that resonated more with Algerians than the non-Algerians. This was an image their brains had already recognized many times before. If mint were a sound instead of a taste, you could say that the French heard the notes; but only the Algerians appreciated the music of it. Because their memories of it were more expansive, mint actually took up more of their brain.

When we are unable to obtain the flavours we remember from childhood, it can give rise to longings so intense it is hard to think of anything else. The anosmia sufferers we met at the start of the chapter such as Marlena Spieler would confirm this: she hankers for the flavours that would make her feel 'like Marlena' again.

Some of the most poignant examples of this flavour-yearning are the food obsessions of prisoners of war. When Primo Levi was imprisoned in a work camp near to Auschwitz called Buna, he remembered that fellow prisoners not only groaned in their sleep, but licked their lips: 'They are dreaming of eating; this is also a collective dream . . . you not only see the food, you feel it in your hands, distinct and concrete, you are aware of its rich and striking smell.'

Among memoirs by POWs of the Second World War, a common theme is not just hunger but the fevered memories it gave rise to of all the things they would eat again once they were free. Very seldom did they build these dreams about the grown-up food of sophisticated restaurants, but the food of childhood and of home: stodgy, filling and safe. One British ex-POW remembered dreaming two nights in a row about 'omelettes and treacle pudding'. He also remembered his bitter disappointment on waking up, since 'Either was as obtainable as a slice of the moon'.[31]

Food obsession reached a particularly feverish pitch among European, American and Australian POWs in the Far East, where the mismatch between their rations of rice and the food they longed for was enough to make them slightly unhinged. Food historian Sue Shephard writes that most of the men in the Japanese camps 'regressed to a childish state'. They all hallucinated about sugar: for the British it might be chocolate eclairs, suet puddings and steaming bowls of buttercup-yellow custard; for the Americans, Hershey bars, mother's apple pie and every kind of layer cake, from devil's food to coconut. Some men refused to join in the collective discussions of food because it was too painful to be reminded of how far they were from home, but for most of them, the crazy food talk became a survival mechanism to get through the endless days of boredom and brutality. A long-term POW recalled that after the first year and a half or so, the food talk had completely supplanted daydreams about women.

Some men went so far as to write down elaborate menus and even recipes on scraps of paper. Film-maker Jan Thompson, who spent twenty years interviewing former American POWs for her 2012 documentary *Never the Same*, found that a common theme was writing down Thanksgiving menus, reconstructed from 'memories of childhood gatherings'.[32] All memory is a distortion, and in their half-starved state, these men constructed holiday menus more lavish than any of them can have actually enjoyed as a child. In Japan, Mess-Sergeant Morris Lewis felt oppressed by the responsibility of looking after his soldiers as well as himself. Sergeant Lewis kept himself 'sane' by writing down an extraordinary Thanksgiving dinner that included Virginia baked ham, fried rabbit, cranberry sauce, snowflake potatoes, candied sweet potatoes, buttered sweetcorn, buttered asparagus tips, green stuffed olives. Then, 'Assorted Cookies', 'Assorted Nuts', 'Assorted Candies', 'Assorted Ice Cream'; also 'Ass. Jams' and 'Fresh Ass. Fruit & Grapes'.

This word 'assorted' is heart-rending, coming from a man whose diet has been reduced to abject monotony. Prison can famously expand the imagination. After all this time without biscuits, nuts, sweets and ice cream in any form, Sergeant Lewis was planning a meal where all these treats are freely offered in multiple varieties. He had returned to that old childhood pipe dream of being given free rein in a sweet shop.

POW yearnings for childhood food were like an exaggerated version of the food nostalgia we all feel. What you are seeking to recover is not just the flavour in itself, but all the things that went with it: your family sitting round the table, the feeling of being cared for, the freedom from responsibility. This is why it's possible to long for bad food too, just because of the happy connotations it may have. Not everyone grows up with a mother who turns out perfect apple pies. POW Russell Braddon, a 'cheeky young Australian gunner' who spent three years in Japanese camps, was thrilled to get a card from his sister. It arrived sixteen months after she first posted it and it had to be short because the limit was twenty-five words: 'Dear Russ, Mum's puddings are still as lumpy as ever. Oodles of love from us all. Pat.' Braddon later said that this letter told me 'all I wanted to know': that his family did not accept he was dead and that 'the old household jokes about my mother's rather abandoned cooking still flourished'.

The childhood foods that we ache for are very specific to the place and the time where we grew up. The American POWs did not dream of 'sweetness' in the abstract but of candied sweet potatoes and pie. E.P. Köster is a Dutch psychologist, an emeritus professor at the University of Utrecht, who works on the knotty question of why we choose some foods and not others. Köster is particularly preoccupied by the role of memory in shaping our gastronomic desires. His career has been unusual in that his work spans the cutting edge of both psychological thought and consumer science. In 2009 he lamented the fact that among many consumer scientists there was such 'lack of understanding' of the 'fundamental insights from psychology'. Köster regrets that consumer research tends to be founded on the assumption that our food choices are rational and conscious when, most of the time, they are anything but.

He traces his own strong preference for dark Bournville chocolate made by Cadbury's – which during the war was the main brand of dark chocolate in Britain – to his memories of being thirteen years old in 1944, when the German-occupied Netherlands was suffering from extensive starvation. One day Köster was out riding his bicycle when a British RAF plane circled overhead. He saw one of the pilots throw a pack containing three bars of Bournville from the cockpit. He quickly grabbed them before anyone could stop him.

> On the way back, I slowly sucked morsels of one bar.
> It was heaven. The other two bars I shared with my
> brother and we ate them for days, a little each day. For
> the rest of my life I have longed for the taste of that
> chocolate and whenever I came to Britain the first thing
> I did was to buy a bar of it. I admit that there may be
> finer chocolates than Cadbury's, but for me there is no
> chocolate more delicious.[33]

Childhood food memories, like family jokes, are often untranslatable to outsiders. If I gave you a small dinner plate containing three mounds, one of cottage cheese, one of chopped apple and one of raisins, you might think me a little odd. You might wrongly suspect me of trying to put you on some kind of low-carb or gluten-free diet. But if I served this to my sister, she would understand at once that I was giving her a nice bedtime snack, just like the ones my mother made for us when we crept downstairs in our pyjamas because we couldn't get to sleep.

The importance of shared childhood food memories for bonding families together can be seen among expats who carry their 'homeland' with them in the form of ingredients smuggled in suitcases. In Greece, they sometimes refer to this desire for the food of home as a 'burning of the lips'.[34] When Greeks move abroad, their mothers will often send care packages of food containing such treats as 'oregano, thyme, mountain tea, locally produced honey, figs, almonds, hard cheese and dried dark bread rings'. At college, I had a Greek friend called Athena, like the goddess, who received the most wonderful parcels from her mother, with slabs of sweet halva and vast bags of the freshest, crunchiest pistachio nuts. She laid them out in exotic pottery dishes. Somehow Athena's student room always felt different from the rest of our, though underneath it was the same scruffy bedsit. Surrounded by her foods from home, she had the air of someone who was never alone.

Until it became easier to buy in supermarkets all over the world, Greeks also travelled with feta cheese. Often, they did not realize how much they would miss this damp white cheese until they were away from home, at which point they became desperate for another salty taste of it. In the words of a Greek academic who got a job at a university in Wales and once travelled back from Greece with a vast 10kg tin of feta cheese: 'I would cut a piece with my meal every night. It was like "white gold" to me.'[35]

One of the functions of traditional cuisines is to reinforce these shared childhood food memories. Food anthropologist David Sutton found that there was a conscious element of remembering in many of the feasts held on the island of Kalymnos in the Aegean. Kalymnians use big ritualized meals as a way of planning to remember events in the future. There is nothing accidental about this remembering. When roast lamb is shared for Easter, it anchors all those eating it in a particular place and time. During a meal, Sutton found that his Kalymnian friends would often say to him, 'Eat, in order to remember Kalymnos.'

These group food memories are a big part of the immigrant experience. As the Italian-American John F. Carafoli has said: 'People start worrying about food and memory when they realise that the food they've loved might not be around forever.'[36] Those arriving in America from elsewhere had an urgent need to buy arborio rice and plum tomatoes; or garam masala and lime pickle, as the case might be. To eat these foods again in the new country was a way of holding on to the grandmothers and mothers who had first cooked with them. Often, however, the remembering through food is bittersweet, because even when you have tracked down every last herb and spice, the missing ingredient is the cook. You find you don't want pasta 'just like Mama used to make'; you actually want Mama herself. After her Hungarian grandparents died, novelist Charlotte Mendelson found herself desperately missing the food her grandmother used to cook for her and her sister: pancakes filled with cream cheese and lemon rind, chicken paprikás and especially 'her meatloaf and stuffed cabbage'.[37] She realized it would be possible, in theory, to find an authentic Hungarian recipe for meatloaf and cabbage and attempt to recreate her grandmother's dishes, but this wasn't the point. 'The internet is no help; I need her.'

At a certain point as a child, we notice that the food at home is not the same as the dinners our friends eat. You discover that some houses smell of burned onions and some smell of spice, while others, disconcertingly, have no smell at all. Even when the same repertoire of meals is served, it is never actually identical to yours. One of my strongest memories – of food, of anything – is being invited to a friend's house when I was about eight. 'I hope you like macaroni cheese,' her mother said, and I nodded, enthusiastically. This was one of our favourite suppers at home. But oh, the disappointment when it arrived. Technically, this was the same dish that we ate: the same elbow-shaped pasta, béchamel sauce and Cheddar cheese baked in the oven until bubbling. Yet everything about it was off. It looked too yellow, it tasted too strong and it smelled of feet – our mother's macaroni cheese was mild and milky. It was like one of those nightmares where your parents are replaced with impostors. This was not the 'macaroni cheese' I knew. It would have been less disturbing if the friend's mother had introduced it to me as some completely different dish rather than passing it off as 'macaroni cheese'.

Given that we all bring such different food memories to the table, how is it ever possible to cook a meal that will please everyone? Chefs are constantly battling with this question of personal memory and how it affects the pleasure we take from a meal. This is where home cooks have an advantage, because we know how different people round the table will react to various foods. We remember the person in the family who will burst into tears if we serve them fish and the one who had a bad reaction once to a mouldy raspberry. But chefs are cooking blind, for strangers whose memories are unknown. The modernist chef Heston Blumenthal once told me that in his dream scenario he would conduct a long interview with every customer at his restaurant the Fat Duck, examining all their deepest food memories – good and bad – before they sat down to eat. Only then would he decide what to cook for them.

Blumenthal himself discovered how radically childhood memories can affect perception of food when he got talking to a friend about tonka beans. These 'black wrinkled beans' are one of Blumenthal's favourite ingredients. To him, the flavour of tonka is somewhere between vanilla, cloves and 'cut hay'. But he had an additional reason to love tonka beans, which was that the smell reminds him of rubbery flip-flops and therefore triggered 'lovely memories of holidays in Cornwall'. Blumenthal was disconcerted to find that when he served his tonka bean sauce to a friend, she hated it, but couldn't explain why. And then it dawned on her. 'She'd had several hefty operations as a child and, she suddenly realised, the bean's rubber aroma reminded her of the mask used to administer the anaesthetic.'[38]

These are the sort of fluke memories no cook could be expected to take into account. But what of more culturally determined differences in food memory? Dr Paul Breslin, a colleague of Mennella and Beauchamp at the Monell Chemical Senses Center in Philadelphia, did an experiment to test the threshold at which people could detect the smell of benzaldehyde, the chemical that gives bitter almond essence its distinctive marzipan smell. This is also the compound that gives cherries and plums much of their flavour. Breslin asked a panel of ten trained tasters to smell different concentrations of benzaldehyde in the air until he found the lowest threshold at which everyone could detect it.[39] He found that nine out of the ten subjects could detect very weak concentrations of benzaldehyde only when they simultaneously drank sugar solution. The sweetness together with the smell reminded them of sweet cherries or plums. The tenth person on the panel happened to be from Japan and responded rather differently. They were the only one who could detect very low levels of benzaldehyde when it was paired with MSG rather than sugar. It must have triggered memories of umeboshi plums, the salted plums that are eaten in Japan with rice. To the Western tasters, the concept of savoury plums did not exist and therefore they did not register it, whereas the Japanese panellist was imprinted with the knowledge of savoury plums from a young age.

How can you find common ground between those who think plums are savoury and those who think they are sweet? Chef Daniel Patterson, the proprietor of Coi restaurant in California (named one of the World's 50 Best Restaurants in 2014), is intensely preoccupied by such questions. He views all food as a form of memory and sees part of his job as being 'expectation fulfilment' while at the same time creating something new. In his book *Coi: Stories and Recipes*, Patterson says he aims at cooking dishes 'with an open, wide-eyed quality, almost child-like in spirit, the way I remember the summers of my youth'.[40] He recognizes, however, that this is a hard task. Patterson knows that when customers from different countries taste the same cucumber, they will all in fact be responding to different things. 'The way an American, a Dane and a Russian understand the taste and context of a cucumber, for example, can be wildly different.' Patterson's job, as he sees it, is to 'make food that somehow triggers primal memories in all of them, using shared experience to create something new'.

He tries to do this by finding a point of 'familiarity'. When diners arrive at Coi, Patterson offers them 'chips and a dip' – in fact, a fancy version of brown rice crackers with an avocado dip – to make them feel welcomed by something they already know. When he serves an ingredient that many diners will not have encountered before and might find alienating, such as duck's tongues, he will pair them with the recognizable crunch of a friendly green salad.

In Patterson's opinion, 'all food is about memory' but perhaps never more so than at the dessert course. His favourite dessert he has ever created at Coi is lime marshmallow with coal-toasted meringue. It consists of home-made ginger marshmallow, blended with lime and frozen in a 'Pacojet beaker' overnight before being topped with meringue and burned with charcoal held with tweezers. The reason he thinks it works so well is that it unleashes 'shared experiences full of emotion': 'it's a childhood memory dish for grown-up palates.' The memory this dish is triggering is the one – universal in American childhoods – of sitting round a campfire toasting marshmallows on sticks. 'I'm not going to recreate a Twinkie,' says Patterson.[41] 'But toasted marshmallows are a cultural experience that Americans share. You go to an uncivilized place, you put a marshmallow on a stick.'

It is ironic that a talented chef such as Daniel Patterson should need to go to such inordinate lengths – the Pacojet, the tweezers, the charcoal, the sharpness of ginger and lime – to replicate the simple memory of sitting by a fire with friends, opening a factory-made packet of marshmallows. But eating is like that. In an age when most of us are, to a greater or lesser extent, reared on packaged foods, few hand-made desserts can equal the emotional potency of a bag of supermarket marshmallows. Patterson needs to find a way of giving you supermarket memories in a restaurant setting. It is not easy. The elaboration involved in Patterson's frozen lime marshmallow shows that recapturing a memory is much harder than capturing it in the first place.

People go on about the first moment that Proust's hero dipped his madeleine in the tea and was transported to childhood. But we hardly ever talk about what happened next:

> I drink a second mouthful, in which I find nothing more than in the first, then a third, which gives me rather less than the second. It is time to stop; the potion is losing its magic. It is plain that the truth I am seeking lies not in the cup but in myself. The drink has called it into being, but does not know it, and can only repeat indefinitely, with a progressive diminution of strength, the same message which I cannot interpret . . .

Whatever was there in the first sip, it's gone by the third. 'The potion is losing its magic.' The surge of memory is short-lived. Scientists call this 'desensitization'. When you try to get back to the past through food, so often you find you can't, because either the food has changed or you have. This is one of the many things that makes packaged foods so alluring. With their bright labels and never-changing fonts, they seem to offer a continuity with the past that you just can't get from other foods. The easiest way to get the hit is to keep going back to the drugstore.

'I bite into my Hostess snowball, and retreat to a world where the only worry is what to ask your mother to put in your lunch box the next day or which pieces of candy you will select at the Kwik-Pak on your way home from school.'[42] Jill McCorkle's vivid description from her 1998 story 'Her Chee-to Heart' captures the nostalgia many of us feel for various processed foods. During the ups and downs of growing up, they kept us company. No matter what horrors or boredom the school day held, reading the breakfast cereal box was a daily thrill, and its contents could be relied on always to taste the same. The names of packaged foods were dreamed up by marketeers with our childish pleasure in mind and formulated to give our sweet tooth a buzz. And when we grow up, we repay the food companies with a loyalty that looks something like filial devotion. When Hostess Twinkies ceased production in November 2012, there was mourning in the US for this highly processed 'Golden Sponge Cake with a Creamy Filling'. Though it was composed of shortening, corn syrup, colourings and other unwholesome ingredients, with a shelf-life so long it became the punchline of many jokes, for many the Twinkie was the taste of childhood. It was Proust's madeleine for the junk-food generation.

When you get three or more adults with nothing in common together, surprisingly often the conversation will turn to the junk foods we knew and loved as children. There is a communal comfort to be had in reciting the names out loud, together, like a liturgy. In Britain, the catechism of nostalgia includes such sweets as Spangles, Jelly Tots, Rolos, Fry's Chocolate Creams, Space Dust. These are common reference points that take us back to some joyous pre-pubertal age when life was free and easy.

No home-cooked food, no matter how delicious, can match the power for bringing people together in misty-eyed recollection of industrially produced food. Convenience foods have been blamed for the breakdown of communal eating. Where families once shared a single dish from a single pot, now they often eat according to individual whim without sharing the experience: multiple microwaveable meals to cater for the taste of each family member or individually wrapped sandwiches and hamburgers. But the experience of junk food is still shared – in the memory, where its emotional force can be alarmingly strong. Unlike traditional food, which is remembered jointly within families or communities, mass-produced food and drink is remembered across continents. How often do you meet someone who hasn't tasted Coca-Cola, at least once? I have no idea what your mother's cooking tasted like, or whether I would have enjoyed her macaroni cheese. But if you tell me she gave you a Mars bar for a treat every Saturday, I'd instantly be able to share your memory. I'd be recollecting the soft squidge of the nougat and the way the layer of chocolate sometimes splintered as I bit into it. Candy bar nostalgia puts us all on the same page.

The power of convenience foods to insinuate themselves into some of our most precious memories – of family, of happiness, of childhood – should be a pressing concern for everyone who is serious about improving anyone's diet, including their own. From babyhood onwards – and maybe before, if our mother ate a bad diet during pregnancy – we are imprinted with memories of junk, like Algerians programmed with memories of mint tea. Our olfactory bulbs have gathered endless sense patterns of foods high in sugar, fat and salt. These flavour memories have become part of the fabric of our sense of self and are not easily discarded, because the system, as we have seen, is designed 'not to forget'. Even when the convenience food no longer tastes good, we still return to it like a homecoming, or like mice pressing the lever to get more pellets, because we remember what the thrill of the dopamine response once felt like.

Mark Bittman, the *New York Times* food writer, asked why it was so hard for us to stop eating foods like hotdogs 'that we know is not only evidently bad for us in the long run but also makes us feel queasy immediately afterward and doesn't even taste good'?[43] His answer was that for him, hotdogs – especially those from Nathan's on Coney Island – had deep memory connections that could not be matched by any of the foods he had learned to love subsequently in adulthood. Hotdogs for him were about childhood and longing and being with his sister at the funfair on a hot summer's day. If Bittman wanted to stop eating them, it wasn't enough to understand rationally that they were unhealthy; or that the meat in them came from the least salubrious parts of unhappy animals. The emotional connection somehow had to be severed.

What makes junk food so dangerous is not that it is unhealthy – though it is. It's that it is entwined in our minds with so many other memories that are good and true and pure. Memory has always been an important part of how we learn to eat, but never before have so many of us been stamped with reinforcing food memories that mostly come not from a cuisine but from a series of cartons and packets. When we hear someone suggesting that we stop eating our favourite brand of ice cream or potato crisps or sliced white bread, we feel a kneejerk hostility. It's hard to let go of these foods and find a better way of eating without a sense of loss. The thing you are losing is your own childhood.

MILK

There is a rising phenomenon of
something called 'continuous milk
intake' in the second year of life. Mothers
arrive at feeding clinics complaining that their
toddler won't eat anything. It often turns out that they
are drinking a litre or more of whole cow's milk a day. This
is an extreme manifestation of a more general phenomenon.
The flavour of sweet milk is perhaps the most firmly imprinted
of all food memories in Western culture. Milk and cookies;
chocolate milk; a mug of warm milk at bedtime: thus we were
comforted, and reassured that we were still children.

Because breast milk is the first thing babies drink, it is
an article of faith for many parents that dairy is the perfect
food for older children too (though anyone with lactose
intolerance would disagree). Some coffee shops now offer
milky 'babyccinos': cappuccino without the coffee, designed
for toddlers. Better a child should drink milk – with its calcium,
vitamins and protein – than ruin their teeth with sugary
fizzy drinks. But our early trust in the goodness of milk has
unintended consequences.

It often doesn't occur to the parents dosing their one-
year-old with limitless milk that anything might be wrong,

because we all 'know' that milk does a child good. In these vast quantities, however, the milk leaves children anaemic (because calcium in the cow's milk blocks the absorption of iron) and badly constipated, not to mention at risk of obesity from the excess calories. The constipation and the fact that the milk is so filling leave them with little appetite for proper meals. As a result, they fail to try new foods, and before long, the sweet, bland taste of milk is almost all they want, because it's all they've ever known.

The problem is not new. In the early twentieth century doctors complained that children were being given so much milk it was making them 'bilious' and overweight. One doctor, Thomas Dutton, complained that such children were like 'drunkards' whose 'thirst for milk is never quenched'.[44] The difference was that at this time milk was expensive and the problem of continuous milk drinking was limited to the 'well-to-do'. Now, industrialized agriculture has made milk so cheap that almost anyone can afford to be a milk drunkard. Even children who are not given excess milk to drink may be weaned onto sugary commercial yoghurts, which also build strong memories of the goodness of milk sugar.

Statistics on global ice-cream consumption (as of 2013, the world ate more than 14 billion litres of it, and rising)[45] suggest that plenty of adults as well as children engage in the near-continuous intake of sweet dairy. And when we speak of drinking coffee, what many of us now mean by that is a 'milk-based coffee drink' such as cappuccino, latte or flat white. A squirt of flavoured syrup takes it closer still to the milk we remember from our toddler days.

CHAPTER 3:

Children's Food

Snack Paks! Before play. Snack Paks! In the park.
Snack Paks! To share.
From a box of Barnum's Animal Crackers[1]

Rice pudding is a childhood food that causes some to shudder and others to shiver with pleasure. It is either a milky bowl of comfort or punishment disguised as dessert. For my husband, one of the benefits of being a grown-up is that he has been liberated from any duty to eat either porridge or rice pudding. When asked to explain the roots of his dismay, the word 'claggy' will feature. If I point out that the texture is just like risotto – which he loves – he responds that no one made him eat that for school dinners. It doesn't matter whether the pudding is baked in the oven until a nutmeggy skin forms or stirred on the stovetop with vanilla pods and lemon zest (my preferred way). The sight and smell make him want to flee. When the children and I eat it, delicious spoonfuls topped with cream and muscovado sugar, he has been known to leave the room.

If we learn to eat mainly as children, then the food we hone our eating skills on is children's food. Yet the education offered by this curious category of cuisine tends to reinforce in various ways the deep-seated belief that healthy food can never be likeable. Over the centuries, the grown-ups who devise children's food have seldom paid much attention to the fact that its composition matters not just in the short term but because it forms the way children will eat in adult life. Modern meals marketed at children send the message that if you are a kid, you cannot be expected to find enjoyment in anything so boring as real, whole food. The kids' foods in supermarkets, laced with sugar and adorned with happy cartoons, teach children that what they eat must be a form of entertainment, portable packages of fun. In earlier times, by contrast, parents and providers did not attempt to make children's food enjoyable, never mind fun, because there was a quasi-religious virtue in being the sort of child who could suffer flavourless food. As the cookery writer Ruth Lowinsky observed in 1931: 'When we were children, it was considered good for our souls as well as our bodies to be continually fed on any food we disliked.'[2]

As we've seen, the main way that anyone learns to like new foods is through repeated exposure. But the example of rice pudding – and of nursery food in general – adds a new condition to this rule. If a food is repeatedly tasted under conditions of coercion or stress, the exposure may have the effect of reinforcing rather than reversing an aversion. For those of a certain age, the problem with rice pudding is that it is what children were *meant* to eat. Growing up, you could not avoid it. 'Cold mutton and rice pudding' is the menu most dreaded by the Bastable children in E. Nesbit's *The Story of the Treasure Seekers* (1899), dreaded because it was inevitable. In the poorer American schools in the early twentieth century, rice pudding was treated as a staple, to be served in many forms. Sometimes it came with bread on the side like soup; in Cincinnati, children were given it in a cone, like ice cream. Milk puddings of every kind – tapioca, sago, semolina, ground rice, whole rice – appeared with such regularity on school dinner menus that the children gloomily eating them might well have imagined that no grown-up ever stopped to think whether it was right to treat a child in this way.

They might have been surprised to learn that for two years in a row at London's Guildhall, in 1912 and 1913, before the First World War took minds elsewhere, some of the top educationalists in Britain earnestly debated rice pudding's role as food for children.[3] When they spoke of rice pudding, what these educationalists were really getting at was the broader problem of children's diets. This is why it is worth returning to this moment in history, to illuminate some of our own confusion about how to feed children.

The 1912–13 debates happened at the end of half a century of intense discussions around the world about children's diets, prompted by growing unease about child hunger. The dominant view at this time was that children – like animals – needed filling up with whatever wholesome fodder their elders chose for them, rather than being given free choice. This dogma started to be challenged, however, by teachers, doctors and social campaigners who reported that poor children's problems with food went beyond starvation. Dr Hall, who worked with slum children in the city of Leeds in the north of England, found that many of them had no clue how to chew food: 'they put it into their mouths,' said Dr Hall, 'and down it went like a letter going in a letter box.'[4] There were reports of children who could not hold a spoon; of two-year-olds who were addicted to pickles and strong cups of tea. What these children needed, suggested Dr Hall and many others, was food that could serve as an education in how to eat.

The period from the mid-nineteenth century to the First World War was a rare moment when improving children's food became a matter of serious politics, thanks to the spread of free and compulsory education. From the 1860s onwards, a school food movement had sprung up to address the problem of large numbers of children arriving at school too hungry to study. How could the authorities force children to attend school during mealtimes without taking some responsibility for feeding them? By 1912, school food had been reformed in Switzerland, Germany, Italy, Denmark, Norway, Sweden and Britain. Paris led the way with the Cantines Scolaires, dining rooms that continue in the city to this day offering good cheap meals to students and school children. At the original French Cantines Scolaires, in 1867, the poorest children could produce a meal ticket and be given a hearty lunch such as roast veal and macaroni cheese or meat broth followed by boiled beef and lentils.[5]

In Britain, the options were far less varied. A new system of school lunches came in with the Provision of Meals Act of 1906, but the mainstay was 'The Eternal Rice Pudding' as the *Daily Mail* called it. In 1912 a Manchester grammar school's dessert menu ran as follows:

Monday: stewed fruit and custard, rice pudding
Tuesday: rice pudding and jam
Wednesday: rice pudding, tapioca pudding with jam
Thursday: rice pudding and stewed fruit
Friday: rice pudding and jam[6]

Rice pudding – which in Edwardian terms was a health food – had much to recommend it as a food for children. It was filling. It was cheap. It was rich in both milk and 'farinaceous' matter, which the best experts in nutrition agreed was wholesome fare for the young.

But did children like it? The 1912–13 rice pudding debates were part of two conferences on the future of school food in Britain which urgently considered the question of how children's health could be improved. This was more than a decade after the Boer War had revealed that many young British men were too undernourished to fight. The stated aim of the conferences was 'rearing an imperial race'. Despite their supposedly nutritious school meals, many British children remained in a shockingly poor physical condition, with widespread tooth decay, stunting and gastric malaise. In some schools, teachers reported that children actually lost weight after the new meals were introduced, because they refused to try them, preferring the stodge they knew.

This was where rice pudding came into the conversation. Delegates stood up and debated the case for and against milk puddings. What they were really debating was whether food should be designed to cater to children's tastes. At this time the prevailing view, inherited from the Victorians, was that children must eat food that was good for them, even if it didn't taste good. *Especially* if it didn't taste good: to eat bland rations without complaining showed that a child had moral fibre. But at the Guildhall in 1912, some made a new and radical proposition: what if the best food for children was whatever they enjoyed eating best?

Several prominent teachers argued that it was time to abolish rice pudding from schools. They did not dispute that rice pudding was healthy, but saw it as a test case for the question of whether children's own tastes should be considered when choosing their food. Their answer – a slightly daring one for post-Edwardian times – was 'Yes'.

Mr W.A. Nicholls, the head teacher of a slum school in Portsmouth, argued that it was a 'specialized sort of cruelty' to force a child to eat rice pudding (it did not occur to him that a child might eat it of their own free will). He confessed that he himself detested rice pudding and never ate it.[7] Mr George Rainey, who ran a children's canteen in London, said that when he served rice pudding to forty poor boys, much went uneaten. In his view, children 'appreciate something which requires masticating' and 'have an aversion to food which is neither liquid nor solid, such as thick soup or rice pudding'. At the opposite end of the class spectrum, Dr Clement Dukes of Rugby School agreed. 'Children crave for sweets,' Dukes remarked, 'and so do I.' Instead of wholesome but insipid milk-based desserts, they should be allowed such treats as jam pudding.

A different view, however, was offered on rice pudding from the delegates who had come to the conference from Bradford, an industrial centre in the north of England. At this time, under the guidance of a visionary medical officer named Ralph Crowley, Bradford was the greatest pioneer of good school food in the country. Before school meals were brought in, the city had suffered some of the worst child malnutrition in Britain. Dr Crowley oversaw a team of doctors who examined all 60,000 school children in Bradford. He declared that more than 6,000, or 11 per cent, were malnourished. The new meals established in 1906 needed to address the children's 'protein starvation' rather than a general lack of food, said Crowley. The Fabian politician Margaret McMillan, who worked with Crowley, said that the great aim of the Bradford meals was to avoid 'stupid feeding'.[8]

Crowley insisted that the lunches must be varied, appealing and above all educative: the meals should teach children hygiene (clean hands and faces!), how to sit and eat calmly, without 'undue noise or bustle' and, most importantly, how to acquire new tastes, instead of the diet of canned food and coffee they were used to. Each day, the children of Bradford were given a two-course meal, rotating on a three-week cycle, with plenty of protein, fat and vegetables and not too much sugar. Crowley was a deeply humane man. When people spoke of the difficulties of feeding children, he replied that there was 'only one point' that mattered, that 'the child must not suffer'.[9] Under Crowley's supervision, the dinner tables in Bradford schools were laid with cloths with a vase of flowers or plant in the centre of each one. Every effort was made to tempt the children to try new dishes. Including rice pudding. Unlike Mr Nicholls, the delegates from Bradford did not consider it a cruelty to serve rice pudding to children.

The difference was that the Bradford envoys were the only ones who seem to have grasped that children's food ought to be devised with a view to a person's future development. They started from the assumption that when the quality of food provided is good enough, children can learn new and more beneficial tastes. A colleague of Crowley's, Miss Marion E. Cuff, who devised the Bradford menus, stood up to defend rice pudding to the assembled company at the Guildhall. 'She noted that, while it might not be a favourite in London, in Bradford "it is the pudding that the children like better than any other".'[10]

As Miss Cuff spoke, it became clear that rice pudding in

Bradford schools was something far more delicious than the dessert served under the same name in London. The kitchen equipment used to make the school meals in Bradford was said to be the finest at any schools in the world, with porcelain baths for washing vegetables and special steam-jacketed boilers. In Bradford, rice pudding was cooked at a very slow heat for three hours with plenty of milk and nutmeg until rich and creamy. The London rice pudding, by contrast, was 'an economical affair', the conference heard: waterlogged and cooked without much care.

Another big difference between rice pudding in Bradford and London was the way it was served. Just as in London, the Bradford children often rejected milk puddings when they first tasted them. But Crowley and Cuff did not take this initial rejection as a sign that rice pudding was a food that children would never like. They understood that some of the poor children arriving at school were used to eating nothing but bread at home and therefore needed gentle encouragement to accustom them to rice pudding. In Bradford, milk puddings were presented in as great a variety as possible – rice, sago, ground rice and others. No child was forced to try any of them, but after a little encouragement, said Miss Cuff, 'they get to like them all'. New foods were offered in small helpings and with individual attention to reluctant eaters. Each table had a 'monitress': an older girl dressed in an apron, who was painstakingly trained to help the younger children eat, without rushing them. The aim was to get the children to the point where they actively relished the 'tissue building' foods that would help them grow.

This suggests real wisdom about children's food. The Bradford school food pioneers saw that children's food did not have to be something healthy that children did not like (such as old-fashioned watery rice pudding) or something unhealthy that children did like (such as jam pudding). With good cooking and a patient but persistent approach at mealtimes, it could be food that was both good for the children and enjoyable. Crowley recognized that his task was the long-term formation of healthy food habits, rather than just filling up 'small savages' with whatever they were used to eating.

This crucial insight has not endured. Even more than adult food, children's food has tended to treat pleasure and health as enemies. Either you are 'eating up' your greens for the sake of obedience and health or you are being indulged with 'naughty' treats. If you internalize these lessons as a child, the odds are you may never fully shrug them off.

In the end, the rice pudding debates went nowhere. The First World War diverted attention away from the finer points of children's taste. For decades afterwards, British school children continued to be served milk puddings of varying quality, with little expectation that they would enjoy them. Chef Rowley Leigh, born in 1950, ate milk pudding 'at least twice a week' throughout childhood 'at home or at school' and recalled that while 'greedy little chaps like me would lick their lips, others would recoil'.[11] So nothing much had changed.

But these discussions in 1912–13 were still remarkable. It wasn't just a matter of rice pudding. Here was a high-level conversation about whether children's food should be likeable or merely wholesome. For most of the children who had come before, it had been neither. One journalist who attended the 1912 conference, from the *Evening News*, was left 'flabbergasted' by all the discussions of exactly what school children should eat. In his own school days he was told 'to eat anything he could jolly well get and be jolly thankful for it'.[12]

There are three basic ways of thinking about children's food. Each of them can teach attitudes to eating that come back to bite you in later life. The first – which I'll call 'family food' – is that, past the milk drinking of babyhood, children's food is no different from any other food: everyone in the house, adults or children, grabs what they can get from the common pot. This teaches you to eat fast and seize your chances when food is available. Family food is the way that children have traditionally eaten – and still do, in most cultures around the world. The second – which we might call 'nursery food' – says that children's food should be separate from adult food, but that grown-ups should carefully select the foods with a view to what they believe is wholesome, rather than catering to a child's tastes. This teaches you that swallowing what you do not like is a form of good behaviour. The third – which I'll call 'kid food' – says that children should be fed exactly what they like, no matter how sugary or fake. This teaches you to satisfy your every whim, and that it's normal to eat bread-crumbed and heavily processed items at every meal.

The ideal form of children's food would take elements from each of these. The best children's food would be as likeable as 'kid food' yet as wholesome as 'nursery food'. And it would also not stray too far from the diet of grown-ups – like 'family food'. The signs are that children eat best when their diet has the same variety of tastes as adult food. But this only works when the family food itself consists of good varied fare: if your parents eat nothing but junk, you might be better off eating a separate nursery diet after all. The best children's food is that where adults control the nutrition but children control what they put in their mouths. Assuming the food on the table is all reasonably nutritious, children should be encouraged to explore the tastes that they like. Which might or might not include rice pudding.

In most places, for most of history, children's food has not existed as a separate category after the age of weaning. If you look at early childcare manuals from the seventeenth and eighteenth centuries, you find that the anxieties are mainly about what a mother should eat to produce healthy breast milk rather than any special foods for babies. In 1662 the herbalist Nicholas Culpeper wrote of 'how to produce milk with a pleasing scent and colour without any sharpness or "ill taste"'.[13] A nursing woman, in Culpeper's view, should have plenty of salads and radishes, moderate amounts of wine, and abstain from fried onions, spiced meat and anger, all of which could make a baby fall sick. But Culpeper said nothing about what a child should eat once they graduated from milk to solid food.

This silence on the question of children's food reflects the fact that protracted breastfeeding was the norm. Aside from a few starchy foods – gruel or sops of bread or overcooked rice in broth – the question of what to feed toddlers was settled by milk. Bone analysis of some of the Medici children buried in the basilica of San Lorenzo in Florence has confirmed that they were milk-fed for nearly two years. By this age, children were ready to eat what the rest of the family ate – with a few exceptions.

Historically, most of the rules about children's food were negative ones. Many societies have had taboos about substances that children must avoid, most often meat. No meat until second dentition – the age of six or seven – was a rule of thumb in some parts of Britain in the eighteenth century.

Among the Mayans of Guatemala there was a belief that all animal foods – whether eggs, milk or meat – were bad for preschool children, a belief that often resulted in them becoming stunted from lack of protein. For the Chaga tribe of Tanzania the meat taboo was more specific. Children were warned not to eat the tongue of the animal, lest it make them quarrelsome; or the head, which could make them stubborn.[14]

Aside from these prohibitions, however, children's food was not generally treated as a special category. Once they had passed the age of weaning, children ate the same food as adults, but less of it, and of worse quality. Children's food could be summed up by the word 'scraps'. In the hierarchy of a working family, children were lower down in the allocation of nutrients – particularly protein – than their father, though they might rank above their mother, depending on how selfless a person she was. There was a harsh logic to this. Without the man's ability to work at heavy manual labour and earn money, no one else would eat.

You can tell a lot about the power dynamic of a family from whether the most highly prized titbits are reserved for the parents or the children. In today's families, preschoolers may be lavished with special meals of organic blueberries and tender chicken fillet while, a few hours later, the exhausted parents make do with something on toast. In former times, by contrast, it was children who put up with oddments, after the parents had taken their portion. In his memoir of growing up in slavery on a plantation in Virginia, Booker T. Washington recalled that in his family, children's meals were ad hoc. 'It was a piece of bread here and a scrap of meat there. It was a cup of milk at one time and some potatoes at another. Sometimes a portion of our family would eat out of the skillet or pot, while someone else would eat from a tin plate held on the knees, and often using nothing but the hands with which to hold the knees.'[15]

Life under slavery was not typical. But the haphazard nature of children's food held true for free labourers too. Mothers gave their children as much food as they could, but only once the man's needs had been taken care of. The British working classes spoke of 'relishes', which were the man's prerogative. Relish in this sense did not mean pickles or condiments, like Gentleman's Relish, the strong anchovy paste marketed to the rich since 1828. The man's relish was what we might consider to be the main course, namely the protein part of the meal: bacon, faggots (a kind of meatball), salted or fried fish, shrimps, beefsteak, or an egg. When these foods were available, they went automatically to the father, both to give flavour to the dullness of his bread and potato diet and to give him the strength to work. Neither children nor women could expect to have 'relishes', unless the father thought to give them a taste. When Dr Ralph Crowley examined the physical condition of children in Bradford schools in 1907, he found that they were suffering not so much from general lack of food as 'protein starvation'.[16]

When trusted with money to buy a midday meal for themselves, children did not necessarily fare much better. Despite being starved of protein, they sought out more cheap carbohydrates. In London, they tended to buy fried snacks on street corners. In New York, at the turn of the twentieth century, the progressive John Spargo (author of *The Bitter Cry of Children*) observed a group of children in a schoolyard going to a delicatessen to spend their lunchtime pennies.[17] Out of fourteen children (eight boys and six girls) seven bought pickles and bread, four bought just pickles, two bought bologna sausage and rye bread and one bought pickled fish and bread. In 1910 a public health campaigner, Louise Stevens Bryant, watched New York school children buying lunches from stores and pushcarts near the school. 'The lunches bought in this way were as follows: a tiny frankfurter and roll, costing one cent; a Swiss cheese sandwich, costing two cents; two small bananas and two long licorice "shoestrings" costing two cents; two frosted cup cakes, costing three cents.' Bryant sent the foods off to a nutrition laboratory to be analysed. The frankfurter was heavily dyed with cerise pink dye. It offered a meagre 5g of protein. The 'lunch' of bananas and licorice gave a mere 0.6g of protein.[18]

All too often, when served family style, children's food failed in the basic task of feeding, which is to nourish. In the 1910s Maud Pember Reeves conducted a four-year investigation into the living conditions of 'respectable' working-class families in Lambeth in London living on 'round about a pound a week'.[19] These were married men, working in such jobs as fish fryers or plumber's mates, not the poorest people in the district. But money was so tight that any protein coming into the house was tightly policed: 'meat is bought for the men,' noted Reeves. One of the families she visited was a carter and his wife, with four children under the age of five. Their usual breakfast was a loaf of bread between the six of them, with an ounce of butter, tea and 'kippers extra for Mr X'. Over the course of the week, these children had little to eat except for bread, tea, potatoes, gravy and greens. There was almost 'no variety'. The occasional appearance of a tomato, when tomatoes were cheap, was a noteworthy event.

The way children eat has always depended a great deal on class and money. Reeves noted that in the households of 'well-to-do' people there were two kinds of diet: one for adults and one for children. In the middle-class nursery, children were given the bland creamy concoctions that were deemed so good for them. In households where the weekly food budget was 10 shillings (£40.62 in modern money, based on purchasing power) the whole family had to eat a single diet, dictated by the needs of the man. Milk was hardly ever purchased, because it was so expensive. It cost the same in Lambeth as in Mayfair. So no rice pudding. 'Nursery food is unknown for the children of the poor, who get only the remains of adult food.'[20]

In all the working-class households Maud Pember Reeves visited, the main food for children was bread.

It is cheap; they like it; it comes into the house ready
cooked; it is always at hand, and needs no plate
and spoon. Spread with a scraping of butter, jam or
margarine, according to the length of purse of the
mother, they never tire of it as long as they are in their
ordinary state of health. They receive it into their hands
and can please themselves as to where and how they eat
it. It makes the sole article in the menu for two meals in
the day.[21]

The food of poor children in the countryside was not necessarily any better. A doctor who lived in the West Country noted that the poorer families there lived on bread and butter, potatoes, 'indigestible' pasties and stale tea.[22] These patterns of feeding children the same food as everyone else in the household were not necessarily thoughtless. Given the high rates of infant mortality, many families believed – rightly or wrongly – that home was the only place where children were safe.

The 'family food' that the children ate may not have been very nutritious, but sitting around the same table, eating the same dishes, gave the family a sense of solidarity. Oxford historian Siân Pooley has studied working-class families in three separate regions of Britain in the nineteenth century, and found that one of the commonest fears some parents had about children was that they would consume food outside the home, particularly fruit.[23] 'Death by fruit' is frequently written in the local court records listing infant deaths. This was partly because it was easier to blame a child's death on a factor away from home than to contemplate the horrible possibility that it could have been caused by anything the parents themselves had done. But the belief in 'death by fruit' went deeper than this. For a long time one thing that everyone, rich and poor, could agree on about children's food was that raw fruit was a dangerous substance for the young.

The fruit phobia probably had its roots in the intensely seasonal nature of fruit. After months of deprivation when no fresh fruit was available, people, especially children, gorged on it straight from the tree when there was a summer glut and made themselves sick. In an era when there was scant knowledge of epidemiology, raw fruit seemed one of the few clear and visible causes of child sickness, something concrete that could explain why so many died in infancy. The fruit phobia also tied in to ancient ideas about the balance of the humours through diet. In Renaissance times, fruit was seen as 'corruptible' and borderline poisonous, especially the sweetest and most tempting varieties such as peaches, sweet grapes and melons.[24] A seventeenth-century book of prayers and songs for mothers and children warns a child not 'to eat much

Of Plums, Peares, Nuts and such
Be they never so Ripe
Lest they thy bellie gripe
Breed in thee ill bloud, cause Collique
Peaches make one melancholique.

Then there was the question of pips. Well into the twentieth century, books on feeding children make much of the need to strain the seeds out of any fruit dishes given to children. Cooked fruit was deemed safer than raw, but safest still was fruit both cooked and deprived of as much fibre as possible. The 'pulp or seeds' of raspberries or strawberries might 'cause serious disturbance' to the child, warned one authority.

Fears that children would harm themselves by eating fruit may sometimes have been justified. The fruit that these children were eating was not a sterile bowl of cut-up melon or a carefully washed apple. Often, children picked fruit from the tree before it was ripe; a basketful of green unripe apricots can certainly cause a nasty stomach ache. Other children might have become ill from picking dirty or contaminated fruit from the ground. Yet the parental fear of fruit – like all fears about food – was not wholly rational. The main reason to be suspicious of fruit as a food for children was that it was just so delicious. It was the candy of its day, something children chose for themselves, away from grown-up control. It was widely noted that children had a special passion for fruit: its tender pulp, its juicy sweetness. Memoirs of childhood often describe berry-picking, returning on a late summer evening inky-fingered and replete. Gathering wild fruits – blackberries, bilberries, blueberries – has always been one of the ways that children could supplement their diet behind adults' backs. Their small size and nimble fingers make children especially good at berrying from low-lying bushes. Families would sometimes use children as berry-pickers to bring in a little extra cash, though they could not always be relied on not to eat half on the way home. In his boyhood, the writer Thoreau used to be sent out to pick black huckleberries for a pudding. He marvelled at the liberty of this wild crop: 'wholesome, bountiful and free'. Grown-ups, however, were not always so happy to contemplate children taking pleasure into their own hands in the kingdom of fruit. Anything so appealing to young palates could surely not be trusted.

During the nineteenth century, a new version of children's food emerged, one that applied the traditional mistrust of fresh fruit to the whole of the adult food supply. The new 'nursery food' of the middle and upper classes was rooted in Victorian notions that children were almost a different species and must be kept physically and morally pure. Nursery food was part of a bigger shift in attitudes to childcare. Family historian Christina Hardyment writes of 'the growing divorce between parent and child' from the 1870s onwards, as parents started to trust the scientific advice of experts over their own instincts. In former generations, middle-class children were brought up in close proximity to their parents, but now they were pushed by nannies in prams, banished to the separate space of the nursery and fed on specialized 'scientific' diets that would supposedly not upset their tiny stomachs.

Nursery food was moral, as much as physical. Medical experts took poor parents to task for their wrongs in failing to give children separate foods. Thomas Dutton MD proclaimed that one of the great 'errors' in feeding children was giving them 'food suitable only for adult people'.[25] In Dutton's experience, the 'majority' of mothers were guilty of it and then they wondered why their children were always ill: '"What do you give your baby to eat?" is a question often addressed . . . "Oh! He eats the same as we do – a little potato and gravy, sucks a bit of meat, and sometimes has a drop of daddie's beer". This is the way thousands of children are brought up.'[26]

Nursery food, with its carefully calculated rations, was a reaction against the rambunctious free-for-all of family food. It aspired to apply rational science to the question of what children should eat for supper. Lying behind these plain, dull dishes – calf's foot jelly, bone soup – was an acute awareness of how vulnerable children were to disease and death. As one Victorian writer on digestion put it, the majority of 'infantile diseases' arose from 'improper food'; the implication was that giving children 'proper food' could save lives.[27]

When it arrived as a separate category in the nineteenth century, nursery food came with its own language. The foods that were approved of – many of them in the rice pudding family – were spoken of in childcare manuals as *wholesome; rational; proper; safe; digestible.* Certain things might '*safely be given*' or '*may be admissible*'. For example, 'Cocoa, if not made too strong, is . . . quite admissible' for children said one cookbook of 1874.[28] Others were *unsuitable; objectionable; unfit; excessive,* causing *biliousness.* As a rule of thumb, the most unsuitable foods were ones that children most wanted to eat, the ones that were rich or sweet or extremely flavoursome ('highly seasoned'). Anything too stimulating in the way of sauces, it was feared, might bring on a bilious attack. Mushrooms, capers, rich tasty gravy, thick cream: none of these was recommended for children. The safest options in the nursery were plain and bland. In Victorian cookbooks, the section on children's cookery, if there was one, usually came next to the one on invalids. Nursery food treated children as if they were permanently on the verge of nervous collapse.

Luther Emmet Holt was dubbed on the cover of his book as the 'foremost authority on babies in America'. His best-selling guide to feeding children went through many editions after it was first published in 1894.[29] Holt's publishers boasted that hundreds of thousands of Americans had been raised on the book and were now using it as a guide to bring up their own children. Simple 'ordinary' food is Holt's mantra, with nothing too tempting to a child's appetite. Running through all his advice is a stern conviction that what is good for adults is not good for children. 'Many foods are useful for adults but too indigestible for children,' he warns.

'Stewed tomatoes,' for example, 'may be given' to children but only once they reach the age of seven or eight. It was those pesky seeds again. Holt believed in sieving all vegetables until the child was three, and continuing to mash them with a fork until the age of seven or eight. Most omelettes, likewise, were 'objectionable' before the age of seven. Eggs should only be given if lightly coddled, poached or boiled, never fried. Nor was Holt an advocate for rich and preserved meats such as 'Ham, sausage, pork, liver, kidney, game and dried and salted meats and fish; all of these are best withheld until the child has passed the tenth year'. Still more dangerous was salad, which, being 'somewhat difficult to digest' should be avoided until the child was eleven.[30]

Most dangerous of all in Holt's book was any kind of pudding, pastry or tart, especially those involving jams, syrups, nuts and dried fruit. Some said that a little sweet stuff would do no harm, but Holt disagreed because a little was 'very apt soon to become a great deal'. The only kind of desserts that Holt could put his faith in were 'junket, plain rice, cornstarch or farina pudding without raisins, baked custard'. A moderate portion of ice cream might be given once a week. Expressly forbidden, however, were: 'All fresh bread and rolls, buckwheat and other griddle cakes, waffles, all fresh sweet cake, especially if covered with icing and containing dried fruits. Lady fingers, plain cookies, ginger snaps are about as far as it is wise to go with children up to seven or eight years old.'

Holt was not alone in this fear of children eating fresh baked goods. It was often said by experts on nursery food that children must never be given fresh baker's bread. The rationale – as with fresh fruit – was that it was both too tempting and too difficult to digest. Two-day-old bread was deemed safe but if the bread contained currants it might be safer still to wait a full eight days. Ideally, it would be 'staled' still further in the oven until it was crisp enough to give the teeth something to work on.

Nursery food came in two textures: very hard and very soft. On the one hand, most of the 'safest' foods were the consistency of mush ('goodnight mush,' writes Margaret Wise Brown in her iconic 1947 storybook about nursery bedtime, *Goodnight Moon*). The idea was to make things soft enough so a child could eat them with a spoon. Oatmeal, bread and milk and custard puddings were all *admissible*. Vegetables for children must always be stewed so soft they could 'pulp through a cullender' as one expert said.[31] Much sieving went on before a food could be reckoned harmless for a child's delicate stomach. Meat should be pounded and not cooked too hard (in earlier times, nurses pre-chewed pieces of meat in their own mouths before offering it to a child). Cereals and grains must be cooked until a sticky gluey mass. Legumes such as peas, beans and lentils were sometimes deemed valuable, because of their high protein content, but only if they were boiled and thoroughly sieved. And even then, there was a fear that they might not be easily 'digested'.

Behind these two words 'digestible' and 'indigestible' lay a world of anxiety about a child's toileting arrangements and what these might mean for the prospects of a child living or dying. Milk pudding was digestible; tomatoes were indigestible. Before the nineteenth century it was considered healthy for substances to move freely around the body. In the pre-modern mindset of purgatives and leeches, diarrhoea in children was not a cause for concern; many saw it as a sign that the body was correcting itself. By the 1890s, however, diarrhoea and sickness were finally recognized as worrying symptoms in very young children and a neurosis sprang up about any food that might be too 'opening' in its effects.[32] The fear of the consequences of gastric upset in children was well-founded, but it led the proponents of nursery food into the realm of paranoia about anything remotely fibrous.

In 1909 Eric Pritchard, a British doctor, expressed terror about the 'intestinal trouble' that could result from allowing a child to eat marmalade, on account of the orange peel.[33] He also warned in the strongest terms against spinach which he noted, slightly surprisingly, was 'a highly popular vegetable in the nursery'. It was Pritchard's finding that 'If the stools of children be examined after meals containing spinach, practically the whole of the spinach will be discovered in a completely undigested state.' Today's baby-rearing books will sometimes warn, in slightly jocular terms, of what you may find in a child's nappy after they have eaten sweetcorn, but there is never any suggestion that the child will be harmed by it. For the proponents of nursery food, though, there was peril to a child in any food that passed too quickly through the digestive tract.

All this nursery mush kept children in a state of constant babyhood. The endless sieved vegetables and slippery milk puddings were not dissimilar to the panadas and paps that formed a baby's first solid food. At the same time, there was a view that children must be given plenty of very hard food – crisp stale toast and the like – in order that they might learn to exercise their jaw and teeth. Great emphasis was placed on mastication. The child whose food did not teach them to masticate was at risk of many 'evils' from stomach complaints to adenoids. Dr Wallace of the London Hospital noted that most problems of digestion arose from bad teeth and that it was therefore very important to include plenty of 'mouth-cleansing' foods in the child's diet such as crusts and toast and dry rusks.[34] As with nursery food in general, the function of these jaw-exercising foods was to do the child good, rather than give pleasure.

The food writer Elizabeth David, who was born in 1913 to an upper middle-class family, recalled the desperate boredom of nursery food in the 1920s. 'We ate a lot of mutton and beef plainly cooked, with plain vegetables,' she remembered.[35] There were 'odious puddings' of ground rice or tapioca 'invented apparently solely to torment children'. She 'hated' the boiled watery vegetables she was given: 'green turnip tops, spinach, Jerusalem artichokes, parsnips'. Everything that David was served in her nursery – devised by her mother 'in league with Nanny' – was designed to be nourishing. She wasn't expected to like 'the obligatory mugs of milk'. That wasn't the point of children's food.

Looking back, it does seem an eccentric way to treat

children: to feed them fare that no adult would countenance eating if they could avoid it. The Italian food writer Angelo Pellegrini complained about having to eat such 'awful stuff' as a boy: watery pieces of polenta dipped in 'foul and evil-smelling' preserved pilchards.[36] Pellegrini's grandfather 'sought to console [him] on such occasions' by telling him that he too had had to eat polenta and pilchard when he was a boy, and that it was even worse for him, since the pilchard was hung on a string above the table and re-used from meal to meal. During the nursery school years, the awfulness of children's food was something that each generation had to endure, before making their own children suffer as they once had.

Perhaps there are still traces of the nursery food mentality around. In some families, the injunction to avoid salt in a child's diet for the first year spills over into a generalized avoidance of flavour, as if a ten-month-old couldn't handle the pungency of garlic or paprika. To dine with the parents of toddlers may entail a meal of plain boiled broccoli and plain roast chicken with no gravy, no salt or pepper and everything separate. A surprising number of people – even those who themselves enjoy complex flavours – worry that no child will eat a plate of pasta unless it is simply dressed with plain butter. For the most part, though, when plain meals are doled out now, the reason is not to thwart a child's appetites but to satisfy them.

The past fifty years or so have seen a near-total transformation in our definition of children's food in the West and, increasingly, elsewhere too. The rice pudding years are far behind us. Once wartime scarcity was finally over, the food supply rapidly industrialized and an array of new convenience products aimed at children appeared on the shelves, which bore little resemblance to the old family staples. Each decade after the war saw innovations in the children's food sector. Hot milk puddings gave way to cold sweetened yoghurts in individual plastic tubs. Fish – as of 1953 – came frozen and ready to cook as day-glo orange 'fingers'. Pies segued into Pop-Tarts (launched in 1963), slabs of jam-filled pastry that a child could put into the toaster for themselves after school. Potatoes were reinvented as waffles and sweet waffles were jazzed up with chocolate chips. Whipped cream became squirtable. And then, so did cheese.

Where in the past manufacturers aimed their messages at the parents who bought the food, they now found that there was money in aiming products directly at the children. Somehow, a new generation of children were able to manipulate their parents into buying them exactly the foods they desired, which were the ones they saw advertised on TV. The fact that many parents complied was a sign of how attitudes to child-rearing had changed, with a shrugging off of old-fashioned wartime attitudes and rising numbers of women working outside the home. The new babycare bible in Britain was *Your Baby and Child* by Penelope Leach, first published in 1977. Leach believed – and it was a liberating attitude in many ways – that the answer to better parenting was 'fun'. Where Dr Spock told parents not to keep fizzy drinks in the house and insisted that children should only snack on fruit, Leach was relaxed about commercial snack foods. The 'lowly potato crisp', she insisted, was 'a surprisingly good source of vegetable protein'. In Leach's view, it was unfair to say that snack foods were 'all rubbish': 'A hot dog, for instance, is a nicely balanced item of diet. Dairy ice cream from a reputable manufacturer is an excellent food, and at least as good for your child as a homemade custard or milk pudding.'[37]

Leach thus absolved her readers of any pangs of remorse they might feel about buying the new convenience foods aimed at children rather than cooking them a homespun dinner from scratch.

Since the 1950s, children's food has gone from being something nourishing but pleasureless to something whose prime aim is to pander to childish tastes. Marmalade and spinach are still not considered ideal food for children, but, unlike Dr Pritchard in 1909, it's not because we fear they will do terrible things to their insides but because we lack the imagination to think that a child might actually enjoy the ferrous taste of spinach or the bitter peel of oranges. The 'kid food' of modern times is designed to please; and so, given the reinforcing power of positive exposure, it does please. 'Kid food' is based on the presumption that children have a natural palate for simple carbohydrates, fat, sugar and not much else. As we've seen, there is no truth to the idea that children have an innate drive that will automatically make them like hamburgers more than grilled fish or muffins more than fresh berries. But if you eat enough 'kid food' meals as a child, the presumption of limited tastes may become a self-fulfilling prophecy.

'Kids' menus are all about fun food,' said a report on children's food in US chain restaurants in 2001.[38] In other words, 'Don't expect any spinach or broccoli.' A journalist in the hospitality business trawled a database of kids' menus from the top 500 American chains. As you'd expect, French fries were a common occurrence. Just *how* common is more surprising. Of the nearly 2,000 menu options at these restaurants, 710 were fries. They were on menus more than twice as often as any other single item, paired with anything from hotdogs to spaghetti. If you were a child in 2001 and your parents suggested a meal out, you could be pretty sure you would be able to order fries.

For your main course, they'd probably let you have something else deep-fried: over half of the 'entrees' sampled were fried, and the rest were usually burgers or pasta. Commonest were breaded and deep-fried pieces of chicken, dolled up in various guises to make them more 'kid-friendly': 'strips, tenders, bites, nuggets, chunks – even antlers'. Desserts tended to be ice cream, often with added confectionery. The Ragazzi's chain served 'Dirt for Dessert: chocolate pudding with chocolate chips, whipped cream, and a gummy worm' (only $.99 – bargain!).[39]

Restaurant meals – as a special outing – may not be typical of what a child eats the rest of the time. After all, who doesn't like to be sitting somewhere away from home eating something hot, crispy and fried once in a while? I often order tempura or crispy fried squid when eating out, though I hardly ever make them for myself in my own kitchen. For many children, however, the fries – and the ice cream and gummy worms – of a typical chain restaurant meal were of a piece with the food of daily life. The top three school lunches in Britain in 2000 were pizza; burgers; chips.[40] 'What does he like to eat?' I asked the mother of one of my son's friends when we arranged a playdate some time around 2005. 'Oh, you know, normal kid food,' was her reply. This turned out to mean chicken nuggets, oven chips, plain pasta, ketchup. No vegetables.

The entire thrust of postwar children's commercial food was to make it seem 'normal' for a child never to eat anything nutritious. Parents who grew up eating rice pudding were not going to do the same to their own offspring. Children's food products were designed to be as fun and stimulating as toys. Even potatoes had smiley faces.

There have always been children's foods that play around with shape and colour. In the past, however, they tended to fall into the category of an occasional treat: a slippery black rope of liquorice shoestrings, a packet of fizzy Love Hearts. During the postwar years, however, the big players in the food industry started to see that they could lavish their visual creativity on children's products that were – supposedly – to be eaten as actual meals. Much of the food supply now resembles sweets, both in nutritional content and form.

Samira Kawash, the author of *Candy: A Century of Panic and Pleasure* (2013), observes that parents have become confused on the subject of sweets. There is a hysteria now in many circles about children eating actual sweets, things like jellybeans that are mostly sugar and colouring. There is a 'nebulous feeling that candy may be dangerous, perhaps even deadly', says Kawash.[41] We know that letting our children eat too many sweets makes us a bad parent, hence the pointless ritual at Halloween when parents allow their children to go from house to house accumulating a big haul of treats, only to confiscate them at the end of the night, because they don't want their child to get cavities. Yet despite their anxiety about sweets, parents will happily feed their children highly sweetened sports bars, fruit snacks and cereals which are sweets in all but name. Why is a bowl of frosted cereal loops with added rainbow marshmallows allowed to count as 'breakfast' and not 'sweets'?

Now, foods marketed for children come in a cornucopia of shapes. Like medieval gingerbread, chicken nuggets have assumed many forms: dinosaurs, giraffes, spaceships, elephants, numbers; and Buzz Lightyear. The old 1960s stand-by of alphabetti spaghetti has been joined by canned pasta shaped like the Teletubbies; Barbie; or Spiderman. And oh, the cereals! The sugar-coated, cocoa-laced spheres and flakes and pops in happy pop-art packages. A market report on children's food boasted that 'developments in extrusion technology' were making an 'increasing range of shapes and textures possible' for children's cereal. Likewise, by the mid-1990s, traditional potato crisps were losing market share to 'extruded' snack products which could be given more 'child appeal' by being shaped like teddy bears or ghosts.[42]

While the form of 'kid food' is more varied than ever, however, the content is far less so. Foods marketed specifically at children tend to be higher than average in salt, sugar and fat. If you want to find an extra-sugary breakfast cereal, pick one aimed at children. As of 2000 several kids' cereals for sale were more than 50 per cent refined sugar by weight. A 2013 study of 577 food ads aimed at children found that nearly three-quarters of them were promoting foods of 'low nutritional quality' despite the fact that more than half of them also included some kind of health message.[43]

Something strange was going on here. Marketeers spoke of a new trend for 'entertaining and amusing a child with food'.[44] Children have always liked to play with food. Maybe you used to pull a croissant apart and pretend the ends were devil's horns; or maybe you took a bunch of cherries and hung them from your ears like earrings; or used tangerine peel to give yourself vampire teeth. Another fun game was using ketchup to dye mashed potato different shades of red, swirling it in with the tines of a fork. I'd add to this the pleasure of eating French beans by opening them up one by one, discovering the specks inside, like green pearls.

The difference with the new kids' foods marketed from

the 1990s onwards was that whereas in the past playing with food felt a little bit subversive, now the games had already been decided for the children by the manufacturers. You were meant to play and the rules had been set up in advance. New children's foods were twistable, stringable or dunkable.[45] There were cheese strings that invited the child to dismember a strip of processed cheese into finer threads and 'dunking' products that included both biscuits and cheese sauce in the pack.[46] Such products were created not with consideration for what the child's body needed but after extensive market research into what children wanted. It doesn't take a genius to figure out that children are not going to tell focus groups that they want more broccoli and rice pudding. Consumer panels showed that children craved products that were just 'for them'.[47] They wanted bright cartoonish colours, smooth textures, sweet tastes. They wanted foods that – unlike the old family food – did not need to be shared with anyone else. Manufacturers responded with products such as dessert in a tube that could be torn open and squirted straight into the mouth or yoghurt with a corner of sprinkles tucked into the lid.

Then there were 'Lunchables' (launched by Kraft in 1988), efficiently contained meals in plastic trays that treated children as if they were passengers on a cramped long-haul flight, far from the nearest supply of fresh produce. In 2002, a typical 'Lunchable' consisted of three separate compartments featuring minuscule hotdogs ('no need to heat'), three tiny white rolls, some cheese food slices ('a good source of calcium') and a sachet of ketchup.[48] This was supposedly a complete and balanced lunch for a child, and one that required no adult assistance to *eat*. What children really wanted was to be treated as older than they were. A market researcher who worked with an average of 4,000 children every year, found that their single greatest wish was for 'control'.[49] The more a product could answer their aspiration to be treated as more grown up than they were, the greater its chances of success. This wish for autonomy over the food supply partly explains the success of breakfast cereal as a children's food. 'Even the simple act of pouring out a bowl of cereal and adding milk gives the child control,' the market researcher noted.[50] Likewise, ketchup became a children's beloved food partly because it is one of the few elements in a meal a child can add themselves.

As of the mid-90s, 77 per cent of French four- to seven-year-olds had the power to choose which breakfast cereal the household bought and 58 per cent could choose their own yoghurts.[51] And this is France, where – or so we imagine – parents still have a tighter grip on family cuisine. But how can wholesome nursery food compete with hundreds of new and heavily advertised concoctions, calculated to appeal to a child's sense of novelty? The labels have plenty of messages designed to assuage any sense of parental guilt. There are 'paediatrician-approved' sweet biscuits and dentist-recommended sugar-free fruit squash, not to mention the endless 'calcium' declarations on sugary yoghurts and processed cheeses that can make you feel you are actually neglecting your child's health by failing to buy them lurid orange cheese-like slices.

The sense that children need their own special foods that are uniquely appealing and altogether different from a mainstream human diet – like pet food – starts early, with commercial baby food. It is easy for anxious new parents to get the sense that they are doing the right thing by turning to packets and jars to feed their growing baby, rather than mashing up home-cooked food. A survey of 5,000 British mothers found that only 35 per cent had offered the baby anything they had prepared themselves on the previous day.[52] Eighty-two per cent had offered food from jars, which, despite the various nutritional claims on the label, is likely to be far less nutritious than home-cooked purées. An analysis of 'fortified weaning food' showed that it was less rich in vitamins and minerals than the old nursery-food staple of sieved potato with an egg yolk stirred in.[53] When choosing what to feed a pre-verbal baby, parents cannot pretend that they are being controlled by 'pester power'. But they may still have a sense of being pestered by the dream babies on the baby-food boxes, who look so pink-cheeked and contented to be eating their apple and strawberry dessert.

Parents tell focus groups that one of the reasons they give in to 'pester power' is cost. Even when a child is not actually there with you, sitting in the supermarket trolley, grabbing things and going red in the face with anger if you do not buy the Thomas the Tank Engine fromage frais now, now, NOW, there is the worry that if you don't buy foods with child appeal, you will make expensive mistakes that go uneaten in the cupboard. A US research company, Langbourne Rust, followed mothers as they went grocery shopping and found that even children as young as one could influence what was bought. Parents would deny their child's request for specific foods only one in three times. This tallies with the experience of Dr Keith Williams, director of the Penn State Hershey Children's Hospital Feeding Program. 'While it should be the case that "children eat what their parents serve",' says Dr Williams, 'our clinical experience tells us that "parents serve what their children eat".'[54]

It is by no means true that all parents now feed their children on 'kid food'. The past decade has seen a modest back-lash against the unhealthiest children's menus. A 2009 survey unearthed growing numbers of vegetables – not all of them fries – on US kids' menus.[55] Even McDonald's now serves organic carrot sticks on its kids' menus. Thanks to the efforts of Jamie Oliver, in 2005, school meals in Britain were re-formed again, with Turkey Twizzlers and similar forms of shaped processed meats eliminated from the menu. In the US Michelle Obama promoted healthier diets for children through her 'Let's Move' programme. In both Britain and the United States, the introduction of healthier school meals was controversial, as many of the new fruits and vegetables went from 'tray to trash' untouched, as one article put it, as chil-dren who knew only 'kid food' rejected unfamiliar dishes.[56] Some took this rejection as a sign that children are natur-ally inclined to prefer 'kid food' to wholesome home-cooked meals. The real lesson, however, is that for dietary reform to be effective, it must go hand in hand with changes in the way that individuals learn to eat. A child will only benefit from a healthy balanced lunch when he or she has developed a taste for healthy balanced food.

The effort of avoiding all the junk marketed under the um-brella of children's food, coupled with the epidemic of child-hood food allergies, has driven some affluent parents a little unhinged. There are households now where children's food is policed more fiercely than it was by L. Emmet Holt in the early twentieth century: where kale is given as a snack and sugar is an absolute 'no-no' and anything containing white flour is treated as only slightly less suspect than hard drugs. The journalist Zoe Williams describes 'wholemeal parents' who refer to raisins as 'baby-crack, to underline their impos-sibly delicious, contraband nature'.[57] In alarming times, food can seem like a way to keep your child safe from danger, and admittedly there are good grounds for thinking that children are at risk from the current food environment.

But the way to protect them in such an environment is not to keep them in a bubble where all the foods are nutritionally perfect. What children need is to develop the skills to navigate the environment for themselves. The problem with this purist version of children's food, just as much as the unhealthy kid food, is: What happens when they grow up? All 'children's food' comes with an assumption that one day you will stop being a child and eat something else.

People from those parts of the world that have still not completely given in to the Western diet speak of how strange they find the concept of 'children's food' altogether. Baby food is one thing. Contrary to what people often say, babies in India are not weaned onto highly spiced foods. During the first year, they might be fed various bland mushes of well-boiled vegetables, with added ghee to boost the calories; or overcooked milky cereals. Suji kheer is thought a good food for babies. It is semolina cooked with sugar and milk – in other words, not dissimilar to rice pudding. After a child is one or so, however, they move on to roughly the same variety of foods that the rest of the family is eating, in both texture and taste, though extra efforts are made to give the child enough protein. In India children's food is just food. Depending on what family you have the luck to be born into, it might be good food or bad food; it might be enough or not enough. The crucial difference with this set-up is that the food of childhood is not something you ever have to outgrow. In the West we think we will outgrow it. But the truth is we often don't.

*

During the Second World War, the American anthropologist Margaret Mead was executive secretary of the National Research Council's Committee on Food Habits. One of the questions Mead was asked to look into was the puzzle of how people could be made to change their food habits.[58] The background was the worry about how Americans might be encouraged to accept the deprivation of wartime food shortages, particularly of meat. Mead saw that people did actually very often change their food behaviours. The pitfall was that when a diet felt constricting, people were liable to swing in the opposite direction as soon as they were free to do so. She gave the example of childhood. Families brought up children to eat less meat, drink more milk and eat more vegetables than their parents:

> Generation after generation, children are reared with the recognition that the customary diet contains foods, some more and some less approved, and they are exhorted to make the choice of foods which are 'good for you' as a matter of moral choice. At the same time, there is implicit in the forms of persuasion and reward, an expectation that most children as they grow older, especially male children as they become men, will insist on making willful choices in favor of foods which are not good for them.[59]

As Mead recognized, the concept of nursery food was based on a double standard. If it was true that there was one stage of life when it was vital not to ingest anything unwholesome, it might follow that at a later stage these unwholesome things might suddenly be allowed or even encouraged. The most blatant example is alcohol. In many families, it was a rite of passage for fathers to get their sons drunk. By the same token, once you were a man, you could safely eat steak and shun green vegetables without anyone telling you off. If anything, it only added to your manliness. It showed that you were no Mummy's boy.

For girls, too, the rules of food changed on the cusp of adulthood. Elizabeth David, the food writer, recalled the wonderful moment she was free to leave the nursery and take tea downstairs in the drawing room with the grown-ups.[60] Suddenly there were elegant sweet cakes and delicious small sandwiches that actually tasted of something. She would never again be made to drink milk or eat rice pudding. It was relatively easy to outgrow 'nursery food' because there was never any expectation that you would like it.

What's less obvious is what happens when people outgrow the unhealthy and highly flavoured 'kid food' of the post-war years. Or whether they do outgrow it. Have you noticed that when someone wants to express that something tastes extra specially wonderful, they will often invoke childhood? Ice-cream sundaes are described as 'So good you'll think you were a kid again', which signals not just that the whipped cream is rich and the chocolate sauce is darkly luscious, but that you are allowed eat it without the burden of adult guilt. At David Chang's Momofuku restaurants in New York and Toronto, they sell a dessert called 'cereal milk' that is meant to taste like the milk left behind after a bowl of cornflakes. And it does: malty, milky and sweet. You can opt to have it plain, or frozen as ice cream.

In theory, we should all reach a level of maturity where we put our childish tastes behind us. We swap our sweets habit for a coffee one. Salad becomes a part of our lives and we grow to appreciate bitter flavours: espresso, chicory,

Campari and soda. Desserts are laced with spirits (tiramisu) or flavoured with challenging ingredients such as cardamom, as if to render them child-proof. Many fashionable dinner party dishes involve an ostentatious adoption of once-hated ingredients: chicken-liver crostini, caramelized Brussels sprouts, fennel gratin. This is how it goes for a lucky minority, anyway.

But judging from what we know of the world's diet over the past few decades, it is clear that large numbers of adults as well as children have now become habituated to eating a version of 'kid food' over a whole lifetime: sweet, salty, undemanding to chew and swallow and heavily processed. The kind of menus you typically see at casual chain restaurants suggests that when adults go out to eat they want childish comfort: sweet-salty ribs, breadcrumbed chicken, cheesy pasta.

Professor Barry Popkin has gathered data on dietary change across the world over the past few decades. Popkin found that: 'Globally, our diet is becoming increasingly energy-dense and sweeter. At the same time, higher-fiber foods are being replaced by processed versions. There is enormous variability in eating patterns globally, but the broad themes seem to be retained in most countries.'[61]

This suggests that 'kid food' has given us tastes that are enduringly narrow, increasingly homogenized and very unhealthy. 'Kid food' has a more permanent hold on our tastes than 'nursery food' not because such foods as frosted cereals and cheese strings are objectively any more delicious than rice pudding, but because they are offered without any 'should' attached.

Over the past fifty years, global food tastes have progressively narrowed to what is referred to as an SFS palate:

sugar/fat/salt. No matter what you order in a fast-food restaurant, from hamburgers to salad dressing to apple pie, the odds are it will be united by a common flavour: not sweet-sour but sweet-salty, with an undertow of fat. This matters, because, as we have seen, flavour has a remarkable ability to imprint itself on our memories and therefore to drive our future food choices. Repeated exposure to SFS foods early in life teaches us that this is how all food should taste. This homogenized sweet-saltiness is now ubiquitous in many supposedly adult treats, from pretzel croissants to salted caramel to pulled pork sandwiches.

The legacy of 'nursery food' was to create adults who, for the most part, were only too eager to leave the constraints of the rice pudding years behind them and graduate to something more delicious. The legacy of being reared on 'kid food' may be a state of arrested development when it comes to food. In 2002 a team of researchers devised a five-year study looking at seventy families to determine whether tastes for certain foods remained constant between the ages of around three and eight. Sure enough, in almost every case the children studied continued to like the same foods from three to eight. The truly startling finding, however, was the extent to which the eight-year-olds' tastes matched those of their mothers. Admittedly the mothers had learned to stop disliking some of the foods that traumatized eight-year-olds: these grown-ups were braver about raw onions, for example, and green peas. But the most passionately 'liked' foods of the mothers were exactly the same 'kid food' that the eight-year-olds preferred, and the list looks like a recipe for nutritional disaster. Nearly all of them, adult or child, were most fond of: popcorn, soft white rolls, French fries, chocolate chip cookies, ground beef, hamburgers, doughnuts, processed cheese, pancakes, syrup, muffins, pizza, white sugar. The only food on the list that was obviously healthy was raw apple, liked by sixty-nine children and seventy mothers.[62]

When parents as well as children are eating 'kid food', perhaps it's time to call it something else. 'Kid food' started off as something separate and different from normal food. Now it is close to being the new normal for all age groups. The danger is that when adults have childish tastes too, it becomes very difficult for anyone to break the cycle and learn the pleasures of real food.

BIRTHDAY CAKE

In recent years, something called 'birthday cake' ice cream started popping up. It is a lurid multi-coloured confection, with sprinkles and clots of frosting and chunks of cake running through it. The idea is that it tastes like the lovingly iced cake your mother made for you when you were six and it was your birthday and you gave your friends slices of cake to take home, damply wrapped in party napkins. Except that you are not six any more and it isn't your birthday.

This feels emblematic of how our eating has gone wrong (see also cookie dough ice cream). Birthday cake ice cream is designed to tap into special memories of blowing out your candles and eating a once-a-year treat with your family. But if you can eat it in a cone any day of the year, on a whim, the whole point is lost. The existence of birthday cake ice cream suggests that we can no longer distinguish celebration foods from everyday ones. We are also not too sure whether we are children or adults.

With children's lives full of so many other sweet treats, birthday cake should have lost some of its emotional weight. If anything, though, the stakes are now higher. Birthday cake has morphed into a pure symbol of parental love. Nicola Humble, the author of *Cake*, says that 'I swear each year that

I will not go overboard on my son's birthday cake, and each year the construction is more complicated, more ambitious, more absurd': a treasure chest, a planet with marzipan aliens, a pyramid cake containing a secret tomb.[63]

Birthday cake is one of those childhood foods we find it hardest to let go of. 'Don't they carry one back to all one's parties?' says the heroine of Katherine Mansfield's story 'The Garden Party', looking at a plate of cream puffs. Many a diet is broken because it feels so cruel to reject a colleague's request to share a slice of cake on their birthday. You don't want to be the mean kid at the party who spoiled the games.

It's not birthday cake in itself that is the problem. It's the surrounding culture of food, where sweet treats are ever-present, consumed without ceremony. In France, according to parenting author Pamela Druckerman, home-made yoghurt cake is used as a lesson in delayed gratification. The child helps make the cake in the morning and has to wait until the afternoon to eat it. It's a useful exercise; for adults too. Healthy eating should not preclude the odd cake. But it's good to be able to wait: if not for a whole year, at least for an hour or two.

CHAPTER 4:

Feeding

His outstanding memory about his early life was his mother's urging him to eat, always with a stern invocation: 'Ess, ess, ich sterbe weg' (Eat, eat or I will die).

HILDE BRUCH, 1974 (of a middle-aged doctor in New York City who had suffered a heart attack but could not bring himself to lose weight)

After my parents split up, my father often took me to the station to catch the train back to my mother's house. Although we had usually just eaten lunch, he would offer to buy me a magazine and 'one more thing' to eat before we said goodbye. I could choose anything. I started to see that whatever I asked for during those anxious moments at the station would be granted, even a whole box of shiny Maltesers, chocolate-coated malted milk balls, the sort of indulgence that in earlier days of family life would only have been purchased on rare trips to the cinema or theatre, and carefully shared out between the four of us. Now, the rules of food had changed.

On the train home, as I pored over my copy of *Marie Claire*, the roof of my mouth became sore from sucking the crunchy honeycomb out of each chocolate sphere, with no one telling me to stop.

At first, I saw these transactions at the train station mainly in terms of my own appetite and the happy carb rush I was seeking. 'Mmm, fattening!' in the words of Homer Simpson. Later, it got more complicated. Around the age of sixteen, I became self-conscious about my increasing weight. When the moment came for my father to offer me 'one more thing' a voice in my head would tell me to say I wasn't hungry or to ask for a Diet Coke. But I hardly ever managed it. The prize of the treats was too great. It wasn't just the taste. It was the feeling of merit; if you are being rewarded with food by a grown-up, you must have done something good. For years afterwards, whenever I found myself travelling by train, my automatic reaction was to buy myself some indulgence or other.

It was only much later, as a parent myself, and rather too eager to dole out platefuls of cookies and milk whenever my children had friends round, that I realized the person my father was rewarding was primarily himself. He felt agonized about the divorce. To cast himself in the role of the bringer of treats, the generous provider, could make the goodbyes less painful for him, as well as for me. To give a child the things they love to eat bestows a heroic glow. It feels almost as wonderful as eating. Seeing a child fed reassures you that you have done your duty as a parent, like a mother bird ferrying worms to the nest. During the years following the divorce, my sister was for the most part refusing offers of food, treats or otherwise. The only child remaining to be plied with goodies was me, and I was happy to oblige, beak open.

*

Much of what we learn about eating comes from the way our parents feed us. As a child, you assume that the grown-ups know what they are doing. But they are mostly just figuring things out as best they can, meal by meal, drawing on the convictions and prejudices they inherited from their own upbringing. Some parents use food as a pacifier, to keep the little ones quiet. Some withhold treats for bad behaviour. Others fret about foods that are too rich or too strange for tiny stomachs and pass on a generalized anxiety about eating. Trends in food change from decade to decade, yet our default patterns of eating are largely a response to an older generation's own complicated attitudes to food. Almost every parent wants the best for their child, but they are frequently too hung up on the indignities of the past to see the real problems in front of them or to separate a child's needs from their own urges. A parent who was forced to eat vegetables in an atmosphere of repression may take joy in seeing children enjoy anarchic TV dinners of whatever they please. Likewise, any parent who remembers hunger will have a special stake in seeing a child eat. Feeding, like eating, is a learned behaviour and the methods that most parents absorb for doing it are based on the values of former times when a child needed to be protected from scarcity rather than plenty. Urging a child to take one more bite is no longer the way to keep them safe.

Not all children are overfed. When a child is neglected, one of the surest signs is underfeeding. As many as 5–10 per cent of all children in the United States aged two to five show poor growth related to scant feeding, rather than to some organic cause (such as coeliac disease).[1] The medical term is 'failure to thrive', meaning that an infant is not getting enough food to grow and develop as they should. The single greatest risk factor for failure to thrive across the world is poverty. In those cases, the poor feeding is unavoidable or at least unintentional. But there can be other reasons as well. Feeding is a complex interplay between parent and child and sometimes failure to thrive is caused in part by avoidant eating behaviour in the child, which in turn makes the mother more anxious about feeding. In many cases, however, failure to thrive is a sign that other things have gone deeply wrong in the child's care.[2] One study suggests that up to 80 per cent of mothers whose babies failed to thrive due to non-organic causes experienced a history of abuse in their own past.[3] The fact that a child is not getting enough to eat is often a sign of a home where there is alcohol or drug abuse or domestic violence. In the worst cases, a parent may intentionally withhold food from a child. This is such a chilling thought that it's no wonder we sometimes act as if feeding is the same as love.

Feeding children is an immense responsibility. To take on the burden of someone else's nourishment until they are old enough to do it for themselves is an expensive, thankless and often unwelcome task. In times of scarce resources, having another mouth to feed can entail sacrifices and adjustments from everyone else in the household. (The scandal of baby formula being marketed in the developing world is not just that bottle-feeding is dangerous when water is unclean. It is that it costs families so much. In Bangladesh, a factory worker might spend as much as a third of their income to buy formula for a baby.)

But when food becomes affordable and plentiful, the emotions associated with feeding children start to look rather different. OK, it's still not fun having bowls of spaghetti up-ended, and no chef has to endure the crushing ego blow of having a lovingly cooked casserole spat out as 'too lumpy'. Standing at the toddler swings in the park, parents are constantly swapping gripes about feeding. We bemoan the tedium of carrying 'healthy' snacks around until a bag becomes a mess of rice-cake debris and squashed dried apricots. It maddens us when a child picks at their lunch, then claims to be hungry half an hour later. And the meals never stop coming. Didn't I already do this yesterday? you think when another breakfast has to be produced, so soon after the last one.

In all this companionable griping, however, you hardly ever hear anyone whisper the secret truth that feeding children – especially giving them treats, where the question of likes and dislikes does not weigh so heavily – can be pretty fun. It feels like parental affection in its purest form. Some of us get our kicks at the school gates doling out hugs and pain au chocolat. It's a thrill to see eyes light up on the arrival of a birthday cake or to be the one buying when you visit an ice-cream stand on a hot day. After you've spent a decade or more fantasizing about the candyfloss of childhood – that warm fluffy halo of sweetness – it feels exhilarating to have a reason to buy it again. When something is easy, we say it's like 'taking sweets from a baby'. Actually, the fingers cling on pretty tightly, I've found. The truly easy thing is giving the baby sweets in the first place.

Feeding can be a cheap thrill, something so enjoyable that even children themselves aspire to do it, acting it out for fun. No less than eating, feeding can be a compulsion. To see small creatures gobbling up treats is very rewarding. The pleasure of keeping pets, for example, is largely about bringing home the food the animal likes and watching it vanish, whether it's goldfish flakes sprinkled in a tank or hamster muesli scoffed up and stored in cheek pouches. We often assume that the urge to push food into another mouth is part of a deep maternal (or paternal) instinct and that the act of feeding others is a form of dutiful sacrifice: you are putting someone else's needs before your own. And often it is so. But we still need to account for the fact that feeding is seen as an enjoyable activity by some of the least maternal and self-sacrificing people you could meet. I am talking about video game players.

In 1980 Pac-Man launched in arcades in both Japan and the US and was an instant hit, changing the whole video game industry. It took its name from the Japanese phrase paku-paku, referring to the sound a mouth makes when it opens and shuts. If you play the game for long enough, cherries, strawberries, oranges, apples and grapes appear on the screen. When Pac-Man eats these fruits, they work as a 'power-up', which means that for a while afterwards he scores more points for every dot he consumes. Pac-Man is not eating food in any normal human sense. Nevertheless, the game's basic appeal, which has been replicated by many other games since, is the hypnotic pleasure of seeing a small creature consuming rewards.

Among video game designers, there is a common language of treats. Whether you are playing the game in Mexico City or Moscow, there are items that are instantly recognizable as good things to feed the characters on screen. It might be a hamburger or a hotdog crisscrossed with ketchup, but most likely it's something sweet: a rectangle of chocolate, an ice-cream sundae, a glazed doughnut. 'One thing that everyone knows in every culture in the world is that sweet things are awesome,' says Luis Gigliotti one autumn morning sitting in a coffee shop. Gigliotti is a creative director in the games industry with twenty years of experience designing games for pretty much every platform – from consoles to online games to mobile tablets. I first met him in this same coffee shop when I overheard him having an intense conversation about how 'cool' it would be to see a dog eating peanut butter. Gigliotti, who wears earrings and an LA Dodgers baseball cap, and has heavily tattooed arms, goes by the name of 'Lu' or sometimes 'Dumpsta'. He has worked on games from Grand Theft Auto and Devil May Cry (a 'hack-n-slash' fantasy) to smaller free-to-play 'addictable' games with cute characters and speedy pay-offs.

Gigliotti sees gaming potential everywhere. When he looks at a pastry, he sees a symbol of happiness. 'Why are these displayed on the counter?' he asks, gesticulating at an assortment of sticky buns and croissants. 'They are trying to make us hungry.' Over the years, Gigliotti has developed a clear understanding of what players will accept as treats 'foodwise'. Colour is important, for example. 'Pink's nothing but good,' he says (whether the player is male or female) and red and even blue are appealing too, but not usually green, because we associate it with sickness. This is one reason – albeit not the only one – that you won't often see leafy greens used as a treat in computer games. It is simply not as rewarding for us to see characters eating green spinach as it is when they eat pink cake. Dark brown roasted things, on the other hand, work very well. 'A brown turkey or the side of a beautifully roasted chicken steaming up with all the fixings. It's synonymous with a feast.' But the quickest way to signify 'treat' is to go for something sweet.

Feeding a computer character with treats is – evidently – not the same as feeding a real child. Often, you identify so strongly with the hero, it is as if you are eating the treat your-self. But while neither the food nor the character is real, some of the pay-off is the same. You want to see the treat vanish and the happy face appear and when this happens, you feel great about yourself. The key thing with any game, says Gigliotti, is to create a connection with the main character and their world. The feelings are real, even if the images are not. 'Once you've created empathy, then all the normal things that give you joy or grief will now apply to what happens to that character on screen.'

When I meet Gigliotti, he is working on a new game about a character called Shark Baby. She is an endearing little girl with pigtails but when she sees something she likes, her eyes flip back and she turns into a shark and devours everything in her path, edible or not. It's the player's job to try to placate Shark Baby with treats so that she doesn't wreak too much havoc, to herself or others, on her adventures. In one scene, Shark Baby is trying to cross a dangerous ravine. Her friend, a make-believe tiger, has a magical sack containing 'yummy cakes', which will help her get across to safety. It is our role to take the cake from the sack and use it to guide her over the ravine. By giving her cake, we make ourselves heroic. The emotion we will feel when we do this, suggests Gigliotti, is very similar to the feeling of supporting a charity or helping a stranger.

In his own life, Gigliotti does not subsist on pink cake, nor is he friends with make-believe tigers. He and his wife have a toddler who they feed 'organic everything'. His wife disapproves, he says, of those parents who buy their kids fast food every day. But when designing games, he taps into a different value system. Gigliotti was born in Argentina, to an Italian mother, a very good cook. Money was often tight and meat was a rare luxury. They only moved to the US when he was nine. Once they built their new American life, his mother couldn't understand when her son brought home vegetarian friends. Meat was the thing you aspired to, so why would you wilfully reject it when it was there in front of you?

When he dreams up his games, Luis Gigliotti thinks like his thrifty Italian-Argentine mother, for whom overfeeding was not the kind of problem you could ever imagine having. The reason we want to feed characters with sweet things in games, he tells me, is because they are 'not an everyday thing. Why is dessert at the end of a meal? It's a reward. Sugar is a luxury item. If you are poor, you don't eat sweets. If you can afford sweets, you are living the good life.' He smiles his wide white Californian smile and we look again at the piles of treats on the coffee shop counter, which both of us can afford but neither of us is eating.

The way we reward children with food is based on folk mem-ories of a food supply that has not existed in the West for decades, when white sugar was so rare it seemed to sparkle like snow. Our impulse to make children happy with food is a loving one – sweets for my sweet – and because the motive is generous, it can be hard to spot that what we are doing no longer makes much sense. In Laura Ingalls Wilder's *Little House on the Prairie*, a family friend, Mr Edwards, risks his life to cross a stormy creek to bring Christmas candy to Laura and her sister Mary. These girls only taste sweets once a year and when they see the walking sticks of striped peppermint, and little heart-shaped cakes 'made of pure white flour, and sweetened with white sugar', they are overwhelmed with joy. 'Think of having a cup and a cake and a stick of candy.' Laura cannot resist taking one lick of the peppermint stick. 'But

Mary was not so greedy.' A treat cannot possibly have the same meaning in an era where white-flour cakes are everywhere and candy canes are sold by the dozen for less than the price of a loaf of bread. The modern child has so many forms of sweetness offered to them that, in my experience, a common reaction to a candy cane at Christmas is not joy but mild resentment that the sugar rush is tainted by the minty flavour of toothpaste. The problem now for many parents is not the fear that there will be no Christmas candy but that when the Christmas sweets arrive in the house, the children still won't have finished up the last batch, from Halloween. And yet the loving spirit of Mr Edwards endures.

We still believe we would do 'anything' to give children their treats, even though we probably do not need to swim through stormy creeks to obtain them. There continue to be families where the grown-ups sacrifice their own pleasures to see the children fed. In China, many children are looked after by their grandparents while the parents go out to work. In the cities, the percentage of children being looked after by grandparents may be as high as 50–70 per cent. For the sake of their families, this generation have given up the leisure they might have expected in old age. Without their selfless childcare, many households could not function and the Chinese economy would grind to a halt. There is a Chinese idiom 'Han yi nong sun' ('to play with grandchildren with sweets in mouth') that conjures up a blissful grandparenthood, in which a grandparent relaxes with a sweet while watching a grandchild play. In reality, it is more likely the grandchild eating the sweet, while their elderly relatives toil. A 2009 study from urban Xiamen, a large city on the southeast coast, found that, after a lifetime of work, the older generation were often now working harder than ever: doing laundry, supervising homework, ferrying grandchildren to and from school.[4]

The grandparents are generally the ones responsible for buying and cooking the food. When it comes to their own consumption, the norm is frugality. One Xiamen grandfather told researchers he eats cheap preserved vegetables to save the money for his nine-year-old grandson's education. Grandparents are far less parsimonious, however, when it comes to feeding the grandchildren and the one-child policy means that all the treats in the household go to a single mouth. In 2003–4 a team of public health experts from Sweden and China interviewed the parents and grandparents of children attending four kindergartens in two different districts of Beijing.[5] The grandchildren tended to have the same food preferences as their grandparents, whether for saltiness or sugar, wheat noodles or rice. The quantities they ate were also determined by the grandparents, who expressed 'love and caring' through food. What this meant in many cases was that they drastically overfed the children.

This overfeeding was not accidental; there was a rationale behind it. All the Beijing grandparents interviewed had vivid memories of food shortage and hunger. As we have seen, such memories will inevitably colour a person's long-term relationship with eating. One grandmother said, 'Happiness in life is to eat what you want, to eat the amount you want, and to eat whenever you want to.' Another grandmother spoke of the exquisite pleasure of watching a child eat. 'My granddaughter has always had a very good appetite. She always opens her mouth whenever I feed her. I enjoy feeding her so much.'[6] There was a sense in these families that a child's desire for food must be satisfied at all costs: that it was worth paying a lot of money to buy enough meat to make a child 'strong'. Some brought along snacks every afternoon when they picked up the beloved child from kindergarten. Others used food as a reward for achievements. A prize in a piano competition was, for Grandmother, a reason to buy 'a lot of chips'.

More than half of the children studied – who had an average age of four and a half – were obese. Several of the grandparents expressed the view that it was good to be a heavy child because a fat child will grow up to be strong and tall. This belief was based on their memories of hungrier times. In previous generations, they would not have been wrong to think about fat in this way. Across the world, puppy fat has been seen – and in some places still is seen – as a kind of insurance policy against malnutrition in a growing child. The French had a proverb: '*Pour avoir assez, il faut avoir trop*' – to have enough, you must have too much. In 1912 a British doctor argued that overfeeding a child was far less of a danger than underfeeding, even if it risked 'a certain amount of excess'. Any plumpness would be soon shed during the next growth spurt, for 'while it is only too easy to overfeed the adult, superabundant nourishment is almost impossible in those still growing'.[7]

This was sound reasoning at the time. Those most likely to have survived famine were those who were fattest to begin with. Between 108 BC and AD 1911 there were over 1,800 major famines in China. Then Mao brought fresh waves of hunger. The Beijing grandparents had survived the hungers of the 1960s, and perhaps they remembered being fat children themselves. It is hard for anyone to stay alive in a famine, harder still for a child. 'Children seem literally born to die,' said one observer of the Irish Catholics in Boston in the 1840s, who escaped famine at home only to be met with new shortages in America. Unless parents built up a child's reserves of adipose tissue when food was plentiful after a harvest, the children would struggle to make it through the 'hungry gap' when stores ran low in the spring.[8] In a bad year, adults may lose 5 or 6kg or 50 per cent of their body fat during the 'hungry season' in modern-day Gambia; if a child were to lose that much, they would probably die, unless they had a little extra to begin with. Given that periods of feast and famine were the norm for our ancestors, those of us who came after seem to have inherited what Hales and Barker christened 'thrifty genes', good at conserving fat. We are all descended from survivors, and survivors were the chubby ones.[9]

'At least it means that I treat the child well ... if the child is fat,' said one Chinese grandmother.[10] Another grandmother felt 'sorry' for her own thin granddaughter. This mismatch between a child's weight and what a grandparent sees is not unique to China. The view that a plump child is a healthy child is common among the older generation in many ethnic communities. Baldeesh Rai is a dietician working to persuade Asian families in Britain to adopt healthier diets. As in China, she finds that the cooking for the whole family is often done by the mother-in-law. When Rai gently suggests that a child is overweight, she often encounters resistance from the extended family, for whom it is a good thing – no matter what science or medicine may say – when young ones have adorably chubby cheeks, ripe for pinching.[11]

Among the Beijing families, many parents expressed frustration at not being able to influence how the grandparents were feeding the child. But since they were out at work all day there was little they could do. One mother told her son not to eat sweets, only to be told, 'OK, I will have them when you leave.' A father begged his mother not to feed his daughter so much but was told that she knew how to feed children because she had brought up three of them. In the most extreme scenario, one of the mothers said that she and her husband had decided to move out of her parents-in-law's house. 'This is the only way I can avoid my mother-in-law's overfeeding my child.'

A generation ago, it was rare to be overweight in China. Not any more. The country's dizzying economic growth over the past three decades has been matched with a rapid growth in the weight of its citizens. As of 2010, official statistics indicated that there were 100 million obese people in China, more than five times as many as in 2002. As a percentage of the overall population, obesity is still much less prevalent than in the West. In 2010 only 4.1 per cent of men in China had a BMI over 30kg/m2, compared with 30.3 per cent in Greece and 44.2 per cent in the United States.[12] Yet the worry, as journalist Paul French and statistician Matthew Crabbe explored in their 2010 book *Fat China*, is that rates of obesity are growing much faster than anywhere else in the world, particularly in the cities. Because of China's size, the country now contains a fifth of all the obese people on the planet. 'Famine to gluttony in two generations is quite an achievement,' as French and Crabbe put it.[13]

In some ways, China's obesity crisis looks like a speeded-up version of that of the West, a consequence of changing diets, sedentary lifestyles (the bicycle has made way for the car and a competitive education system leaves many children with no time to exercise) and the modern food industry. Traditional Chinese cuisine – with its peerless balance of flavours and textures – always seemed like a good way to eat, whether for pleasure or health. But in the last thirty years new foods have arrived and, with them, new tastes. Rising incomes for city-dwellers and the opening up of markets mean that they can buy food in both quantities and varieties that would have seemed alien to previous generations. China has developed a penchant for many novel things: fried chicken, supermarkets, beer, frozen foods, burgers, French fries, cook-in sauces, TV dinners, fizzy drinks, breakfast cereals, jam and pizza. Perhaps most surprising has been the success of the big coffee franchises in persuading a nation of tea drinkers (calories in black tea: negligible) to swap to milky coffee (calories in a grande mocha with whipped cream: nearly 400).

Given the influx of these strange new items, it would be easy to blame China's new weight problem on a move away from a traditional diet. 'Don't eat anything your great-grandmother wouldn't recognize as food,' is one of Michael Pollan's rules for healthy eating.[14] A Chinese child eating blueberry muffins and drinking milkshake has definitely broken this rule. But we shouldn't be too quick to attribute China's weight problem to modernity. While many of the foods may be new, the attitudes to feeding are the old peasant ones: if you find yourself in a state of plenty, the right thing to do is to gorge, to set yourself up for the lean times. Among the urban grandparents studied by a group of researchers from the National Center for Women's and Children's Health in Beijing, certain thoughts about feeding were universal: children should ignore whether they were full or not and leave nothing behind in the bowl; waste was anathema; and good behaviour merited treats.[15] In some ways, China's obesity is a symptom of the fact that attitudes to eating have not changed fast enough to adjust to a new situation. Your great-grandmother would not have known how to feed children in these times, because she never encountered such abundance. Like the rest of us, she would have had to muddle through and – with luck – adapt. The Chinese situation is a more extreme version of what is happening all over the world, when a new food supply collides with an outdated body of knowledge about feeding.

All these feeding practices came from a desire to protect the children and see them flourish. These grandparents, survivors of hunger, wanted their descendants to enjoy the things they had never had. Yet the researchers who interviewed them concluded that the grandparents' generous patterns of feeding were pushing their beloved grandchildren to gain weight. These obese preschoolers were not suffering from any lack of affection. They were not neglected. If anything, they were loved too single-mindedly.

What the Chinese case tells us is that we urgently need to invent new models of generosity. We need to find a way for a small portion to feel as much like love as a large one. The longing to protect our children through food – which is one of the most potent forms of affection – now requires new manifestations. 'Comfort me with apples,' says the Song of Solomon. It is such a magnificent feeling, to feed someone with love, no wonder we place great faith in it. This love plays out in all kinds of ways. In an ideal scenario, you love a child by giving them carefully chosen foods that are good for them. It makes you feel cosy, knowing that your child has a thermos of home-made soup on a cold day, particularly if you experience guilt that you won't get home from work in time to make a hot dinner. But expressing parental love through food does not always work so well. A parent's enjoyment in feeding feels so right, we think it will lead us inexorably to a child's true needs, when often it takes us somewhere else.

*

Through the ages children have been traumatized by the injunction to leave a clean plate. For some, it becomes a miserable battle of wills: a child sits for hours in front of a congealing plate of food as a parent or teacher urges the child to eat the hated meal. This is a set-up that can never end well for either party, especially when the child undergoes genuine revulsion for the food forced upon them.

Charles Fourier (1772–1837) grew up in provincial France in the early nineteenth century. A delicate boy with very strong likes and dislikes, his childhood memories were dominated by the 'tyranny' of school teachers and parents in matters of taste. 'How many canings did I not receive for refusing to eat turnips and cabbage, barley, vermicelli, and other moral medicine that caused me to vomit, not to mention my feelings of disgust.'[16] Fourier remembered that when eating with his school teacher, he was often required to eat turnips, which he loathed. One time he tried to throw his turnip away, only to be discovered by the teacher and forced to eat it, now covered in dust.

At home, likewise, Fourier was often forced by his father to finish food that he did not like. On one occasion, his father compelled him to eat leeks, until he was violently sick. The memory left scars. Later, as an adult, Fourier worked as a travelling salesman but in his spare time he wrote books of social theory. The foundation of his philosophy was that no one should ever be forced to do anything that went against their inclinations. He dreamed up a utopia – Harmony – in which instead of bread being the staple food, it would be replaced with a mixture of fruit and sugar, more to children's tastes. For him, forcing someone to eat food they could not stand was a form of child abuse.

Fourier's older sister Lubine, however, remembered things

slightly differently. Charles was the baby of the family – he had four older sisters – and, as Lubine told it, he was his father's favourite. Lubine recalled that their father's strictness when it came to whether Charles finished up his food was a sign of particular fondness. She observed that her brother was 'very delicate about his food'. Her papa's aim in getting him used to 'cleaning his plate' was the thought that '*no one knows the situation in which he may find himself placed in life*'.[17] He bullied him to eat in the hope of toughening him up for a cruel world, since he 'loved him as one loves an only son'. According to Lubine, as soon as he saw his beloved child vomit, he regretted his actions and promised to let the boy 'do as he chose in the matter of food'. But for young Charles it was too late. He would never forget his father's cruelty and he would never be able to eat leeks.

How many parents and children have been locked in such battles, over the years? It starts with a delicate child who is reluctant to eat. This makes the parent anxious that they are not getting enough of the things that are good for them – in this case, turnips and leeks. They try to force matters, which only increases the child's reluctance and turns their suspicion of the healthy food into an active loathing. Whether the food is eaten or not eaten, no one wins in the end.

It does seem a crazy way for parents to behave, but throughout history it has often been driven by a genuine fear of scarcity. For anyone lucky enough to know only postwar plenty, it is hard to reconstruct just how great was the horror of waste to earlier generations. The sight of nutritious food being pushed to the side of a plate is never enjoyable, but in a time of the world wars, and the Depression in between, it began to seem to some like a selfishness akin to crime. In 1940 *The Times* published an editorial calling for it to be made 'an offence' to waste food. My granny was born in 1908 and whenever we ate a baked potato in her house she used to recite this rhyme (half sternly, half in jest):

Dearly Beloved Brethren
Is it Not a Sin
To Eat a Baked Potato
And Throw Away the Skin?
Skins Feed Pigs
Pigs Feed We
Dearly Beloved Brethren
One, Two Three.

Luckily, I thought the skin of a baked potato was the best part, particularly if there was a raft of butter to spread into the crispy crevices. Plus, I was the kind of child who never needed to be coerced to finish. But maybe I'd have enjoyed the skin even more without being told it was sinful not to eat it.

The clean plate – and generally coercing children to eat against their will – is one of a range of traditional feeding techniques, founded on a fear that food shortages are just around the corner. These techniques were impatient ones, devised by parents who did not have time to sit around fussing about likes and dislikes (though, ironically, a determination to wait with a child until they clean their plate can make a meal go on all day). In rural Nigeria it is still the norm for mothers to hand-feed children on diluted, fermented maize called 'eko'.[18] Mothers give the eko from hand to mouth because it is quicker than a spoon, no small matter when the mother works eight hours a day as a market trader. If children resist being hand-fed, mothers resort to force-feeding. Observers have watched as the mother covers the child's nose with a cupped hand so that they cannot breathe, forcing them to swallow the maize porridge.

The assumption underlying such techniques is that a parent knows better than the child what their stomach needs. The psychologist Leann Birch has identified a series of 'traditional feeding practices' all based on protecting children from scarcity. These include:

1. Feeding children frequently
2. Offering large portions
3. Offering food as a first response to crying
4. Coercing children to eat when food is available.[19]

When food is short, such strategies may be a way to protect a growing child. But when obesity is more of a threat than famine, they lose their logic.

Leann Birch has led many experiments whose results suggest that these traditional feeding methods are actively damaging in the modern world, resulting 'in overeating and accelerated weight gain' as well as bad feelings around mealtimes. Feeding children too often can make them forget what their own hunger feels like. Large portions lead to overeating. And giving food to calm a distressed child teaches them that unhappiness is a reason to eat. This last one explains a lot. If your mother interpreted all your cries as a cry for food – rather than for play or sleep or a fresh nappy – then it stands to reason you would feel inclined to treat yourself the same as a grown-up, silencing your sadness with sugar.

As for coercing children to eat up, it teaches many things and none of them is very helpful. If you are disgusted by whatever you are being forced to eat, it makes you fear the food on your plate and the person giving it to you. A study of 140 college students found that their strongest food aversion very often went back to an incident of 'forced consumption'.[20] Even if you are not disgusted, force-feeding trains you to obey the plate and not your own appetite. You learn to measure when to stop eating not by what your body is telling you but by external forces.

It wasn't as if people – at least child-rearing experts – were ever unaware that force-feeding was a bad idea. Manuals on feeding infants and children repeatedly warn against pushing food on a child who does not want it. In 1923 the paediatrician L. Emmet Holt insisted that 'Children should not be continually urged to eat if they are disinclined to do so at their regular hours of feeding or if the appetite is habitually poor, and under no circumstances should a child be forced to eat.'[21] Holt argued that the result of force-feeding was that the child 'has less and less desire for food and may even have attacks of vomiting'. Likewise, a psychological article on feeding from 1944 noted that forcing food and 'too great concern on the part of the adult' about a child's food intake could 'hinder a child's progress in eating'.[22]

But force-feeding can be a tempting strategy, nevertheless. I know, because, to my great shame, I did it to one of my children. L. Emmet Holt is right when he says that it begins 'in despair' because the child's appetite 'is not keen'. Or at least, this was the case for me. My third child was born with a cleft palate, making it harder for him to swallow. When he was tiny, each feed could last an hour or more, at the end of which much of the milk often gushed back up his nose and was lost. Before he had the operation to repair his palate at six months, I fed him through a combination of breast-feeding and expressed milk, which I gave him from bottles with special teats. Even though it felt as if the whole day was dominated by feeding and expressing, he stopped gaining enough weight. The cleft nurses were worried. When I look back now at photos of him aged three or four months, I am startled by how skinny he looks, a little pale skull with wide trusting eyes.

Once he switched to a mix of formula and breast milk, he gained weight again. Following the nurse's advice, we spoon-fed him on purées for two months before the operation. He took to them easily. It seemed to be a relief to him to be eating something thicker than milk. He loved carrots, mashed banana, all sorts of gunk. The operation at six months went fine and now he could swallow like anyone else, thanks to the embroidery of stitches on his palate. The nurses said he could go home as soon as he was eating enough. He learned to eat porridge and broccoli and beef stew and many sorts of dal. He was fine.

It was me that was the problem. Looking back, it's clear I never lost the early anxiety about his feeding. Aged eighteen months, he developed the classic balkiness about food that most toddlers go through. It got worse when we all travelled abroad for ten weeks for my husband's work. He was fussier than his older siblings had ever been. Foods he had once adored he now spat out. At the same time, he became fixated on sweet foods, begging for sugary fromage frais and ginger-bread men. He called these 'Run Runs' because of the story 'The Gingerbread Man': run, run as fast as you can. Maybe he envied the way the gingerbread man could escape. In his own case, it wasn't being eaten he wanted to run away from but being fed. Instead of sitting back and riding things out – as all the baby books, not to mention my common sense would have told me – I started to force matters.

At first, I would make him take 'just one bite', ramming in the tiniest spoonful of something that I 'knew' he liked, saying 'Mmm' in a pantomime voice. To start with, this worked. After the first taste of spaghetti bolognese or whatever, he remembered it was OK and carried on eating by himself. But, increasingly, I found that he was still wildly shaking his head after the first taste. The thought of him skipping a meal was horrible. Maybe he needed another taste, I thought, forcing my spoon between his little clenched teeth. 'Remember! You like carrots!' But he didn't remember.

'I thought you were an awful mother,' said someone at a lunch party who had seen me ramming food into his mouth one day. The more I did it – hardly surprisingly – the more limited his repertoire of foods became. He even began to reject cake. It made me demented to think how little protein he was consuming; and how much sugar. At one stage, all he would consent to eat were bananas, gingerbread men, muffins, dry cereal and yoghurt. These were the only things I wasn't pressing him to eat and therefore the only ones he could safely enjoy. It makes me shudder now to think how offensive it must feel to have a large, powerful person bearing down, cramming a hard spoon between your teeth. 'By far the worst aspect of weaning in my view,' Germaine Greer observed in a lecture from 1989, 'is coming to terms with cold steel.'[23] Pretending that the spoon is a train or an aeroplane only makes it worse: would you like your mouth to be used as a runway? Grown-ups who have been asked to recall what being force-fed was like report emotions such as anger; humiliation; betrayal. Force-feeding is a crime of passion, driven by a parent's desire to see a child eat; as with other crimes of passion, the perpetrator has lost sight of the loved one's autonomy.

The forcer of food always feels they are justified at some level (I know I did). In 2001 a group of students were asked by psychologists to look back on times in their childhood when they were forced to eat food.[24] More than two-thirds of the students had been forced to eat food at least once. In almost every case, the students said that the authority figure's 'stated purpose was to benefit the child'. The most common reasons were to avoid wastefulness; to make a child get more variety in their diet; and to ensure they ate healthy food ('we worry about your health when you don't eat'). In a handful of cases, the reason given was tradition. No fewer than five of the students – who attended Southern Methodist University, near Dallas – had been forced to eat black-eyed peas on New Year's Day, which is a good-luck ritual in that part of the world. One of the students had been called 'un-American' for not wanting to consume a hotdog on 4 July. The most frequently forced substances were vegetables (49.5 per cent) and red meat (15.9 per cent).

In our case, we broke the force-feeding habit. I backed off, and my son slowly broadened his horizons. A few months on, he rediscovered carrots when I steamed some batons and brought them along on a pushchair ride, presenting them as a surprise offering that he could try if he liked, rather than a meal pressed upon him. I stood well back, as if lighting fireworks, and he freely took some in his hand.

You, I am sure, would never behave in such a stupid and cruel way to a child. But there are other, subtler, ways of coercing a child to eat that are far more normal. Interviews with a group of mothers of preschoolers in 2011 found that 86 per cent used 'verbal encouragement' and 54 per cent used 'physical encouragement' to get their child to eat.[25] Verbal encouragement might mean something like saying, 'You can't leave the table until you've finished.' Physical encouragement could be reverting to spoon-feeding a child who no longer needs it or putting food on a fork for them. Another study found that many parents believed that pressurizing and cajoling a child to eat was a positive strategy to use at mealtimes. And indeed it does seem perfectly reasonable. Anyone can see there's a big difference between threatening to cane a child if they don't finish a plate of leeks and pushing them to take three more bites.

Research suggests, however, that even quite mild verbal coercion changes how a child views food. Leann Birch and colleagues set up an experiment over an eleven-week period by getting preschool children in Pennsylvania to eat soup, with and without pressurizing them to eat it.[26] They chose two soups for the experiment: butternut squash and corn. Half the children were pressured to eat only the squash soup and the other half were pressured to eat the corn soup. In the pressure situation, a grown-up calmly reminded the children four times – once every minute – to 'finish your soup'. The no-pressure situation was exactly the same, except that the grown-up made no reminder to finish. Researchers measured the amount of soup the children drank, plus any comments they made. The effect of the pressure varied from child to child. A minority relished the challenge, saying things like, 'Wow. Yellow soup! I think I can drink yellow soup!' and proudly showing the researchers their empty cup. But the vast majority of comments (157 of them) during the pressure condition were negative. Children said things like, 'Yuck, it's yellow soup again,' or, 'I told you already I don't like it.' When told to finish their soup, one child said, 'You always say that to us and I don't want it. It's so annoying.'[27]

Apart from the negative feelings, the children ate a

markedly larger quantity of soup – whichever flavour they were assigned to – when they were not pressured to do so. Over time, the children became less willing to eat the flavour of soup that they associated with being pressured to finish.

In other words, as so often with parenting, the effect of pressure is exactly the opposite of what we hope for. Our desire to see a child eat well, which comes from love, is often so blinding that we cannot see that we are the ones standing in the way of them doing it. The Beijing grandparents thought they would make their grandchildren healthy by feeding them as much as possible. The force-feeders think they can teach a child to be less fussy. In both cases, the strategies backfire.

So what does actually work when it comes to feeding a child?

The question of how a parent's feeding style influences a child's eating habits is so complex that you might think it was beyond the realms of science to measure. Outside a family, no one really knows what goes on at the dinner table. Each household has its own distinctive rules and bizarre unspoken codes. As a child, you come to learn what distinguishes a 'snack' from a 'meal' and whether a request for a second helping will cause your parent pleasure or displeasure. These questions are not easily explained to outsiders, much less quantifiable.

Yet a surprising amount of work has been done on the link between feeding styles and child health and the findings are relatively clear. Some studies have asked children to describe the feeding style of their parents; others have asked the parents themselves to describe how they feed, using a range of inventories and questionnaires. Still others have observed the parents and children at mealtimes. Researchers then tally this data with various health outcomes in the children such as whether they are overweight and how many vegetables they eat. From all these studies, there are strong indications that some methods of feeding have better outcomes than others.

To make things simpler, researchers tend to divide up parenting styles into four main models, based on how responsive or sensitive a parent is to a child's needs (sometimes this is referred to as warmth) and how demanding they are when it comes to the child's behaviour. From these two ideas, we get four basic styles of parenting:

1. Uninvolved: Low warmth and low demands
2. Authoritarian: Low warmth and high demands
3. Indulgent: High warmth and low demands
4. Authoritative: High warmth and high demands

Obviously, there are far more types of parents than just four. It's also the case that parenting styles do not necessarily translate into feeding styles. Some parents have mood swings from meal to meal, being authoritarian at breakfast, before the first coffee of the day has kicked in and indulgent at dinner, mellowed by the first glass of wine. But this model is a start.

Of these, the uninvolved style is clearly the worst. If your parent was an uninvolved feeder, they would have failed to provide the food you needed and also failed to place any demands on you to eat well. We are not talking about the odd dinner of tins foraged from the back of the cupboard. With uninvolved feeding, almost every meal is chaotic and you come home to an erratically stocked refrigerator and parents who do not appear to care one way or another about what you eat. Several longitudinal studies have found that children fed in such a way are predicted to have a higher weight.[28] If no one cares much about feeding you, it's hard to learn healthy ways to feed yourself.

The remaining three styles are less straightforward. The authoritarian style of feeding belongs with the force-feeders. Such a parent makes high demands on a child to eat well (Finish your soup! Try these courgettes! Don't eat sweets!) but fails to recognize the person in front of them and what they really like or need. By contrast, the indulgent care-giver – like the Chinese grandparents – is hyper-aware of a child's wants and food preferences and good at responding to hunger. Where the authoritarian feeder has an arrogant sense that they know best, the indulgent feeder is warm and responsive. Yet an indulgent feeder will place few demands on a child at mealtimes, whether to try new vegetables or to stop eating when they are full. The indulgent feeder will not tell the child to wait until suppertime or that you don't add mayonnaise *and* butter to a sandwich. They will bake every cake – or, more likely, buy them – until they find the one that puts the biggest smile on the child's face.

Some might say that indulgent feeding is a two-way street. In some families, children become skilled at running rings around the grown-ups over food from a very young age. They discover that eating can be a bargaining chip or a way of getting attention. Or they learn that if you ask for something often enough, in whiny-enough voice, you will get the treat you desire. And the treats are everywhere. It is far easier for parents to slip into indulgent feeding than it was in the past, given the sheer prevalence of foods expressly designed for indulging little ones.

Whatever the cause, a growing body of evidence suggests that indulgent feeding puts children at higher risk of obesity.

Rachel Vollmer and Amy Mobley, who conducted a large review of the existing literature up until 2013 on how feeding styles affect child health, found that the indulgent feeding style – however kindly meant – was a strong predictor
of obesity.[29] One study of nearly 400 immigrant Hispanic
families found that the indulgent feeding style accounted
for 26 per cent of the variation in children's weight, even
after controlling for differences in parental BMI.[30] Moreover,
this indulgent style of parenting was the most popular one
among the mothers studied, used by more than a third of
the families. Seven separate studies have found that an indulgent parenting style (sometimes defined as 'permissive') went
along with higher weight in children.[31] Indulgent feeding is
also associated with children consuming more foods low in
important nutrients and high in sugar and fat. This is as you
would expect: indulgent parents indulge their children.

What's perhaps more surprising is that the authoritarian
style of feeding – characterized by placing high demands on
the child to eat well combined with low sensitivity to the
child's own feelings – has also been linked to higher child
weight. One study following nearly a thousand American
children from birth to fifty-four months (four and a half)
found that having authoritarian parents created the highest
risk that children would be overweight out of any of the
parenting styles.[32] Children with authoritarian parents were
five times more likely to be overweight by the time they started
school than those whose parents had a warmer approach. At
least five studies have found a connection between authoritarian parenting and weight gain, though the evidence is
more mixed than it is with indulgent feeding.[33]

In certain contexts, authoritarian feeding may have
some merit. There are signs that with very young children

– particularly in low-income households – placing very high demands to eat fruit and veg and not to eat too much junk can have positive effects, setting up children with healthy habits for life.[34] A dictatorial style of feeding will also have a different impact, depending on the family values surrounding it. One study in New York City found that authoritarian feeding had less negative connotations among a group of Chinese-American families than among a group of white American families. In the Chinese-American families, this parenting style had no impact on the children's weight, whereas in the non-Asian families feeding children in restrictive, controlling ways was associated with higher body weight in the children.[35] There is always the danger that attempts to control a child's feeding too rigidly will backfire.

The great drawback with authoritarian feeding – apart from the unpleasant atmosphere it creates at the dinner table – is that it prevents the child from learning to recognize their own cues for hunger and fullness.[36] Leann Birch conducted another revealing experiment, this time involving 192 girls.[37] Researchers met the girls when they were five and again at seven. They asked the girls' mothers to fill in a questionnaire determining the extent to which they attempted to control their daughter's eating, responding to statements such as 'If I did not guide or regulate my child's eating, she would eat too many junk foods,' with options from 1 (disagree) to 5 (agree). The girls themselves were, after a lunch, given free access to a range of snack foods including pretzels, crisps, popcorn and chocolate chip cookies. Those girls who had the most controlling mothers also tended to be the ones who ate the most food in the absence of hunger. By the age of seven, they were also the most likely to be overweight. Birch concluded that too much restriction promotes behaviours in children that are likely to make them gain weight. We want what we are told we can't have.

Reading these studies as a parent can be a good way to depress yourself. Indulgence makes a child fat. Restriction makes a child fat *and* unhappy. It's tempting to throw your hands in the air and complain, as my mother sometimes used to at moments of stress, 'I know! I do everything wrong!'

There is good news, however, if you have children who are not yet fully grown. It is possible to learn how to feed children in ways that are beneficial rather than the reverse. The tricky part is that doing so involves giving up many of the pay-offs that we get from feeding. You have to relinquish, or at least cut down on, the joy of indulgence – the bliss of seeing a small creature gobble up treats like Pac-Man. You also have to give up the illusion of power or the feeling that the child's stomach is just an extension of your own: the belief that you can make a child stop or start eating because you know best.

The style of feeding that has been shown to offer the best outcomes for children's health in numerous studies is called 'authoritative'. Such parents are highly 'demanding' that a child should eat well. Yet they are also highly 'responsive' to cues from the child (so no force-feeding or coercion). Another way of putting it is 'high warmth, high control'. The ideal scenario for a child is to grow up in a house where there isn't too much junk food, but also not too much fuss made about the evils of sugar and fat. Four large studies of American families have found that where children are fed in an 'authoritative' style, they eat more fruit and vegetables, even during teenagerdom. They consume more dairy products but fewer sweets and sugary fizzy drinks. They eat more family meals and fewer snacks scavenged from fast-food places. They are less likely to be overweight. Perhaps most significantly of all, these lucky children have a lower chance of growing up to be emotional eaters.[38] In 2009 researchers in one study interviewed 450 mothers and their children (with an average age of seven). Children whose mothers were authoritative – warm but in control – were less likely to say that they turned to food when angry.[39]

The end-goal of feeding a child is very different from what it appears to be in the crazy mess and rush of a mealtime, when parents are desperate just to get them fed and cleaned up and on to the next stage. The true objective is independence: for a child to reach the point where they can regulate their own intake of food and to choose the things that will do them good while giving them pleasure. Weaning them off milk is one thing. But the real task for a parent is to wean children off needing you. The dietitian and family therapist Ellyn Satter talks of the 'division of responsibility' in feeding. From toddlerhood to adolescence, a parent should be responsible for 'what, when, where'. The child is responsible for 'how much and whether'. Satter's idea is that over time, a child offered good family meals with the freedom to eat as much or as little as they need will grow up to become a 'competent eater'.[40]

The 'authoritative' model also points to ways in which adults could learn to feed themselves better. So many of us feed ourselves in ways that are alternately neglectful and overly strict. Next time you sit down to eat, imagine you are an ideal parent feeding a beloved child. Wouldn't it be nice if you could offer yourself food in a warm, structured, no-fuss kind of way? You wouldn't punish yourself with crash diets but nor would you allow yourself too much junk. Your priority when choosing food would be to see yourself well nourished and you'd choose meals to keep your mood on an even keel. You'd want yourself to enjoy eating. The pantry would be stocked with good food and you would trust yourself to choose wisely from its contents.

The art of feeding, it turns out, is not about pushing 'one more bite' into someone's mouth, however healthy the food. Nor is it about authoritarian demands to abstain from all treats. It is about creating a mealtime environment where – as in Clara Davis's feeding experiment – those eating are free to develop their own tastes, because all the choices on the table are real, whole food.

A new movement in child feeding has emerged suggesting that children can be taught to 'self-regulate' from a much earlier age than previously assumed. Gill Rapley was a British midwife and health visitor who felt dissatisfied with the conventional wisdom that babies should be introduced to food via purées, given from a spoon. Rapley pioneered a new system called Baby Led Weaning, designed to be introduced at six months.[41] Instead of feeding babies with all those elaborate organic purées frozen in ice-cube trays, you simply place chunks of food in front of the baby: steamed vegetables, soft pear; later, pieces of toast or even lamb chops. All food becomes 'finger food', even messy things like risotto. The baby will either grab it and attempt to eat – or not. Should they decide not to eat it, the parent is not to 'step in' with a spoon. The idea is to mimic the conditions of breastfeeding, when a baby takes as much or as little as they need.

Doubtless, there have always been some parents who fed babies this way. But as an official movement, Baby Led Weaning was launched only a decade ago, as a consequence of the new advice from the World Health Organization to wait until six months before offering a baby solid food. At four months, a baby is not ready to grab food for themselves, but by six months many of them are. This being so, Rapley argues that it is 'logical' to skip spoon-feeding altogether and allow babies to learn eating by 'doing things for themselves'.[42]

The radical thinking behind Baby Led Weaning is that in developmental terms we actually learn deliberate swallowing only after we have learned to chew. Babies, says Rapley, do not have the ability to move food to the back of their throats until after they have acquired the ability to chew. Rapley's innovation was to let babies learn chewing first. Many of the first meals of a BLW-fed baby will provoke gagging and spitting, but Rapley does not view this as a bad thing. When a parsnip is regurgitated, it is not cause for despair but a sign that babies are exploring different flavours and textures. As for the fear that BLW will lead to choking, Rapley claims this is not so, suggesting that there is good reason to think that babies 'are at less risk of choking when they control what goes into their mouths'.

In the decade since it was launched, BLW has found rapturous followers among middle-class mothers, who say it has taken the stress and worry out of feeding. I've spoken to people who say it makes mealtimes a more equal interaction between parent and child, who sit side by side as companions at the table. But, as with the breastfeeding versus bottle-feeding debates, the BLW discussions on parenting websites can quickly descend into rancour on both sides, with the more dogmatic advocates of BLW suggesting that spoon-feeding in any form is akin to retarding a child's development.

What's most striking about the BLW literature is how it takes away the parent's traditional role as feeder. No longer is the parent to view themselves as a beneficent hander out of treats, as the mother bird passing worms to the nest. Parents, says Rapley, must resist the temptation to 'help' a baby who is struggling to get food into their mouth. Nor should they decide when it is time for a baby to stop drinking milk or how many feeds to give. Rapley recommends continuing to breastfeed alongside the solid meals of finger food because a breastfed baby is 'always in control' of feeding and will take more or less milk depending on how thirsty they are. 'There is no need for [a baby's] mother to make these decisions for him.'[43] She must not hurry him or chivvy him. And put down your knife. 'There is no need to cut food into mouth-sized pieces.' Other than buying and cooking the food – which could be anything but not fast food, ready meals or choking hazards like nuts – the parent's main role is to watch.

The evidence so far on BLW is not all positive. On the plus side, the mothers using BLW are required to behave in ways that are less restrictive and pressuring than those using traditional spoon-feeding. Babies fed in this way are more likely to eat the same foods as the rest of the family, which is a bonus assuming that the family eats healthily. But in one small study of American mothers using BLW, it was found that the grown-ups – and by implication the babies – were eating excessive amounts of sugar and salt and inadequate amounts of micronutrients, especially folate.[44]

There are also concerns as to whether all babies are developmentally ready for grabbing chunks of food at six months. Professor Charlotte Wright, a paediatrician from Glasgow, found that (of a sample of 600 families) only 40 per cent of babies were in fact ready to self-feed at six months.[45] By eight months, 90 per cent were ready to reach out spontaneously for food. This indicated to Wright that it was 'unrealistic' to expect children to rely exclusively on self-feeding when they started on their first tastes of solids.

Another problem with BLW is that, as we saw in Chapter One, when a child waits until six months to start solids, they miss out on much of the crucial flavour window between four and seven months, when they would have been more receptive to acquiring new tastes.

BLW cannot be the one true way to feed a child, because nothing ever is. But it does suggest that babies need far less help with feeding than parents generally think, especially when so much of that 'help' turns out to be counterproductive. Our current food environment requires children to learn very different skills from those taught by the traditional feeding methods. It can take years of reconditioning to see that it is not in our interests to take 'one more bite' or clean our plates when we are already full; to realize that the people who taught us to eat were themselves not necessarily masters of the art of feeding. I know a grown man, the youngest of four, whose mother was still feeding him from baby jars when he started school. He is now middle-aged and still has nursery food tendencies. Sooner or later, and preferably sooner, we all have to liberate ourselves from the parental spoon.

LUNCHBOX

The artist Rich Gold once said that the lunchbox could be thought of as a '"portable shrine of home" that is carried into the hostile environment of school'.[46] It is a strong protective enclosure, adorned with beloved cartoon characters such as Spiderman or Hello Kitty and filled with food that makes you feel your parents are close by.

We probably place too much faith in the feeling of security that comes from carrying a lunchbox. The problem with lunchboxes is the problem with the way we feed children in general. Parents trust that anything they place in this magical box will be good for the child, because it comes with their love. A report on school meals in the UK in 2013 noted that just 1 per cent of packed lunches sent from home fulfilled the nutritional requirements applied to the hot lunches served in the school canteen.[47] Yet most parents, the report noted, still believe packed lunch is healthier.

The contents of the average lunchbox offer a pretty strange education in what 'lunch' is. A sample of 1,314 school lunches eaten at American elementary schools in 2014 revealed that packed lunches contained more sugar, more calories, more processed snacks, more sugary drinks and less protein, fibre and calcium than school lunches.[48]

This poorly balanced selection is eaten in unsociable isolation, except for the odd piece of bartering ('Your cereal bar for my chocolate?'). Parents may give their child a packed lunch because they believe them to be too fussy to eat anything else but a 2009 poll suggested that children eating canteen food were far more open to trying new things than the lunchbox kids.[49]

But this magical box still has potential. When it comes to adults feeding themselves, a lunchbox can become a tool for training yourself to eat better. The Japanese bento – pioneered using aluminium boxes in the early twentieth century – offers a structure ideally designed for eating a healthy lunch. Rectangular compartments are filled with varied flavours, artistically arranged: rice, vegetables, protein (tofu or fish, stir-fried chicken, some meatballs or a Japanese omelette) and beautiful fruit. The boxes – which are used for lunch for everyone from toddlers to businessmen – give an easy handle on portion sizes, without any need to count calories. Makiko Itoh, author of *Just Bento*, used bento lunches to lose more than 30lb, because the box focused her attention on

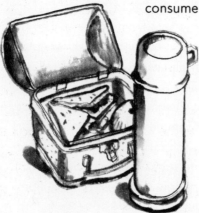

'variety and portion size'.[50] If you only consume what is in your bento – 300ml for little kids, 600ml for adults with modest appetites and 900ml for hearty eaters – it should be impossible to overeat.

CHAPTER 5:

Brothers and Sisters

And now the best food was cooked for poor
Hansel, but Gretel got nothing but crab-shells.
THE BROTHERS GRIMM, 'Hansel and Gretel'

On our first family holiday abroad, a week in Brittany in the north of France, my older sister discovered moules marinières. She would pick out the plump mussels one by one, noting how each one was slightly different, dipping her baguette into the rich winey juices. Two years her junior, I couldn't allow myself to like the same thing as her, so I chose winkles for my favourite food of the holiday. Never tried winkles? You are not missing much. They are sometimes described as 'edible sea snails', but this is to stretch the meaning of the word 'edible'. Their small black shells make them a pretty addition to a French plateau de fruits de mer but they have the texture of gristle and the flavour of salty tears. Every lunch of that holiday, I doggedly ate these cold chewy molluscs – each one as joyless as the last – loudly insisting they were far superior to her fragrant bowl of mussels.

Siblings have always marked out territory through food. Squabbling over who gets the best bits of a family dinner is the main way people first learn to compete over resources. Sharing out a pizza is a tough lesson in justice and how it needs to be blind: you cut and I choose. Eating with siblings provides an early warning in how everyone feels entitled to more. It isn't just about getting the last chicken wing or the last cherry in the bowl; it's about winning. You want to be the person wily enough to grab an extra pancake before anyone else does. My youngest, aged five, can be reduced to weeping by the discovery that his brother, ten years older and unattainably ahead of him in everything, has finished his favourite packet of cereal. The trembling look on his face says: thwarted again. He doesn't mind half as much when the cereal is eaten up by his dad.

As time goes on, the sibling rivalry over food gets more complicated. Or it did for me and my sister. At first it was about who had most: more licks of cake mixture, more scrapings from the edge of the dish of cheese soufflé. Later, the competition became stranger and more devious. I always knew she was clever, because teachers would tell me so when I joined their class two years later and proved less brilliant. But I only realized just how clever she was the day of the custard tart incident. We both adored custard tarts: the kind sold in every British bakery sprinkled with a thick skin of nutmeg with crimped edges. The game here was to eat the tarts as slowly as possible. We started with the dry pastry around the edges and worked our way down to the delicious wobbly cream and last of all, the soggy disc of pastry at the bottom, the trophy – like the nutty heart at the centre of an artichoke. My sister usually won the race to finish last, because my greed would get the better of me. One day, we were each given our tarts and I began to nibble slowly as usual. My sister left the room, came back and said that she had finished her tart. Elated to have won at last, I quickly gobbled mine up in just a few mouthfuls. At which, she produced her pristine tart from the other room and ate it very slowly in front of me.

In the end, competing over the same things got so tense that we started to divide up the world into things that were mine and things that were hers. What she had, I wouldn't have. She got a rabbit; I got a guinea pig. She did art; I did music. She read all the great children's books, such as *Ballet Shoes* and *The Railway Children*; I was left with comics and Enid Blyton. She had mussels; I had winkles. She turned vegetarian; I loved roast dinners. She nibbled lentils and nut cutlets; I ate our mother's beef stew and parsley dumplings (here, in my opinion, I got the better deal; vegetarian food in the 1980s was not the flavour-packed feast it has now become). For years she was famous in the family for how fast she could eat. One of her talents – when not eating custard tarts – was for finishing everything on her plate before the last person had been served. I dawdled over my meals, savouring second helpings; and thirds. She had no interest in cooking, which created an opening for me to play around in the kitchen, making cheese straws and huge puffy brioches, which I would then try to persuade her to eat. Mostly, she declined.

With the teenage years, the games changed, and I would sometimes look back fondly on the old silly battles. It's sad to look at a tray of warm scones that no one is fighting over. She started to miss meals – just breakfast at first, but later stretching into lunch and dinner too, subsisting mainly on green apples. She'd say she wasn't hungry and didn't want to come down from her room, and our parents seemed to accept that. I would look at my mother's worried expression and the empty place mat and I would eat all the more. Sometimes she did join us, but only picked at what was on her plate. I was happy to finish whatever she left. The more she starved, the more I binged. It was like the rabbit and the guinea pig all over again. We were just playing our allotted roles, the only way we knew how.

The power siblings have over our eating habits is no small thing. Yet we hardly ever talk about these familial influences. Whether you develop the habit of eating breakfast or not may be less a matter of temperament than how many brothers or sisters you had as a teenager: data from one study suggests that the more older siblings an adolescent has, the lower the teen's chances of breakfasting, regardless of family income.[1] Having more children in the home makes for a more chaotic morning start, plus, as my youngest has discovered, older siblings have an annoying habit of hoovering up all the cereal.

Even those who grow up with no siblings may assume quasi-brother or -sister roles when they sit down to eat in groups. It's like when someone says 'Shall I be Mother?' as they pour tea from a teapot. In practice, what eating as brothers and sisters often means is that we expect people to receive different shares of what's on the table depending on gender. Across different cultures and countries, we grow up with entrenched – if not fully articulated – beliefs that boys and girls merit separate foods and differing quantities of those foods. Often, these beliefs lead us to feed ourselves and our children in ways that are damaging to our health.

The influence of the companions who joined us around the childhood table continues long after you have stopped eating together. Decades later, you are in a sandwich shop and your decision about whether to have tuna or turkey – or to walk out of the shop and not have a sandwich at all – is partly shaped by this person whose spoon and cup were laid out next to yours night after night. Young children – aged around two to eight – are, according to research, far more likely to resemble their siblings in their food preferences than they are to resemble their parents.[2] This rings true. Parents tend to offer children the same food, whereas they may save certain dishes for themselves, especially in families where children and parents eat in separate sittings. A brother and sister sitting eating meatloaf and frozen peas side by side are modelling to each other that this is a good thing to eat. Each bite that you see the other person take reinforces your liking. Or not: it is hard to sit calmly by and carry on eating if you share a table with someone who is grumbling that peas are 'gross' and pinging them at you with a knife.

As siblings get older, their eating habits become more distinct. When 415 pairs of Dutch siblings aged thirteen to sixteen were followed for a year in 2002–3, it was found that their eating behaviour was only 'moderately similar'.[3] Much depended, as you'd expect, on the quality of the relationships. Brothers and sisters who were very close to each other – reporting having a lot of 'fun' together – had more similar eating habits than those who were distant or hostile. The really startling finding was that in the cases where the siblings copied each other's eating habits, it was the older ones who copied the younger rather than the other way round, especially with pairs of girls. This is the opposite of what you'd expect. Shouldn't the older one set an example? Not in our topsy-turvy culture. The researchers put it down to the fact that the older ones, having gone through puberty, are not as thin. With their teenage curves, the older girls looked at their skinny-thighed sisters and subliminally concluded that eating like a little girl could reverse the effects of time. Most disturbingly, the older girls copied the younger ones when they engaged in 'emotional eating', eating in response to stress, anger or fear. We assume that thin people are the ones who have sussed out how to eat, even if they are only thin because they haven't yet reached a certain age.

It is through our siblings that we first learn how unfair our genetic inheritance can be: that one person seems to be able to eat whatever they like without putting on weight while another is forever struggling to consume less. Inequality among siblings goes all the way to the gut: we are born with different microbes inside us, outnumbering our cells ten to one. Some of them affect our chances of becoming obese in later life and others affect how well we digest our dinner. What we eat is constantly changing the composition of our microbiota, but, equally, the nature of the microbes inside us determines how well we respond to the food we eat. This variation exists even among identical twins. Researchers in Malawi discovered that when a pair of twins with severe malnutrition and kwashiakor – a protein deficiency – are given the same emergency aid rations, one twin may recover while the other doesn't, the reason being different microbes in the gut.[4] Scientists took faecal samples from some of these 'discordant' twins and transplanted them into mice. The mice with the healthy twin's microbiota easily flourished even on a low-calorie diet whereas the mice with the kwashiakor-sufferer's microbes lost alarming amounts of weight.

So siblings are not born equal; but the inequality is often entrenched by the differing treatment we receive at the table from our parents.

Mealtimes are a vehicle for favouritism (or perceived favouritism, since most parents believe themselves to be scrupulously impartial): who gets served first? Who gets the florets of broccoli and who gets the stalks? Who is expected to put up with the slice of burned toast because they 'won't mind'? Who gets the special drinking glass? Whose turn is it to clear the dishes? These parental decisions often feel unfair to at least one if not all of the children involved. In some contexts, for some children, they are positively toxic.

This food favouritism is especially stark when it is done because you were born the 'wrong' sex. The idea that boys deserve bigger or better portions than girls plays out in many ways, some subtle and some not so subtle. In its most extreme forms, this sex bias in feeding has damaging consequences that last well beyond childhood.

If you must be born in rural India, especially the south, don't be born a girl. An Indian girl under the age of five is 75 per cent more likely to die than an Indian boy: 'the worst gender differential in child mortality for any country in the world', reported the *Times of India* in 2012.[5] Sex-selective abortions have led to an artificial gender imbalance in India. In 1901 there were 103 males for every 100 females across the population. A hundred years later, it is 107.2 males for every 100 females.[6] The situation – which extends to many other parts of Asia, too – has been described as 'gendercide'.

For the surviving girls, life is often worse than it is for boys. In many of the very poorest families, girls are given less of everything: less food, less medicine, less clothing. They are not seen as having the same economic and social usefulness as a boy, and therefore parents invest less in them, down to the rice on their plate: not in every family, certainly, but enough for it to be measured statistically. As long ago as the 1901 Indian census, this nutritional neglect of girls was observed:

'She is less warmly clad . . . she is probably not so well fed as a boy would be and, when ill, her parents are not likely to make the same strenuous efforts to ensure her recovery.'[7]

But not all girls are equally deprived. If you must be born a girl in rural India, take care to have older brothers and not sisters. In 2003 the Indian economist Rohini Pande set out to quantify the effects that siblings had on how badly girls were fed.[8] She measured this by looking at the numbers who were 'severely stunted', in other words with a height more than three standard deviations below what would be expected for the age. This is a pretty good marker of a child who is suffering long-term nutritional neglect.

Pande looked at data from 14,715 rural children aged six to forty-seven months. Most of the mothers were illiterate and very few of them earned any money to supplement the father's income. A third of the children lived in households that owned no consumer goods at all. These are families who have very little indeed, where brutal choices are made with the scant food the household has on any given day. In these Indian villages, no one, even the favoured children, has what we would consider 'enough'. Half of these children lived in villages with no all-weather road.

Just how hungry the children are in these families depends on what siblings they have and of what sex. Overall, 6 per cent more girls than boys were severely stunted, which accords with the picture of 'gendercide'. But Pande found that some girls had better chances of being adequately fed than others. Girls who had several older brothers actually had a lower chance of severe stunting than a boy with two or more older brothers. The worst-off girls were those with multiple older sisters. The existence of these other girls took away any novelty value they might have had and left them with the least food of anyone in the family. A girl with two or more older sisters had a 38 per cent chance of being severely stunted, reflecting how little her parents valued her existence. Boys, on the other hand, benefited from being the only male among sisters. Pande found that such a boy had the lowest odds of stunting and the highest odds of being immunized against disease of any children, because as the only boy among girls he was extra special: extra worthy of food.[9]

Such grim choices are – thankfully – far removed from our own dinner tables where there is usually too much food to go around rather than not enough. When tables are laden – whether in India or the West – the family dynamic is no longer about who deserves to starve the most. Here, the unlucky child may be the one forced to play the role of the human dustbin, to make everyone else feel better about the leftovers they do not want to finish. We are certainly not counting out every grain of rice.

Yet the harsh parental choices described by Pande sound oddly familiar. My hunch is that this is because sibling competition for food is a central feature of the fairy stories we still read to get our children to sleep, or the Disney films we snuggle up and watch together on the sofa on rainy afternoons. We know that the Disney Snow White is a good kind person because she shares out seven portions equally between the dwarves' seven little bowls, with no holding back or favouritism. *Cinderella*, meanwhile, etches on our minds the thought of families where blood siblings are given more of everything than a neglected stepsister. In the original story, *Ashputtel*, the ball that she is prevented from attending is described as a 'feast'. By stopping her from going, the suggestion is that the dreadful stepsisters are preventing the heroine from eating as well as dancing.

The harrowing dilemma of how to feed multiple children in tough times is one of the main themes of the Grimms' Fairy Tales, first published in 1812 as *Children's and Household Tales*. Much of the wickedness in the stories centres on stepmothers who are unwilling to give stepchildren as much food as they give themselves or their own children, if they have any. 'Hansel and Gretel' begins in a time of 'great famine'. The stepmother is not truly bloodthirsty – like the witch in the gingerbread house – but very selfish and scared of dying. She wants to abandon the children in the forest because she fears that otherwise 'all four of us will starve to death'.

In the conventional Freudian interpretation, wicked stepsisters and stepmothers are a fantasy. For the Freudians, a woman who outright starves her stepchildren could not possibly be real; she must be a nightmarish projection of the inner anxieties that children feel about being abandoned. Yet in a groundbreaking article published in 1981, the French historian Eugen Weber argued that fairy tales reflected the truth of life in eighteenth-century Europe: 'hunger, poverty, death, danger, fear, chance'.[10] To the peasants who first shared these folk tales, the idea that you might be deprived of bread because you were valued less than a stepsibling was horribly real. 'Given the mortality rates, especially of women in childbed,' writes Weber, 'wicked stepmothers were not a subject of fantasy any more than castout children.'

Viewed in this light, the experience of children in fairy tales is not unlike that of the two young siblings in the extraordinary Studio Ghibli anime film *Grave of the Fireflies* (1988) depicting the famine in Japan during the Second World War. When their mother dies, the brother and sister Seita and Setsuko are forced to try to survive on their own. An aunt takes them in, but she makes them feel unworthy of food at every mealtime, even though they have given her every scrap of food they had. The aunt resents them for not being old enough to work, like her own child. She blames them for not contributing to the household or to the national war effort. But we feel that the real root of her resentment over food is that these poor orphans are not quite closely related enough to her. The scientist J.S. Haldane famously declared that, 'I would lay down my life for two brothers or eight cousins.' In a famine, being vaguely related to the person who controls the food supply may not be enough.

Fairy tales are full of unlucky stepchildren being forced to live off mouldy crusts. Sometimes – as in 'Hansel and Gretel' – the deprivation forces siblings to become allies, to band together to find a better life. In Grimms' 'Brothers and Sisters' a brother takes his sister by the hand and says that since their stepmother beats them and feeds them 'hard leftover crusts' worse than the morsels the dog gets, they must 'go out into the world together'. When fairy-tale children seek their fortune, food is the main thing they are after.

Weber observes that when characters in fairy tales are granted three wishes, they tend to have modest ambitions: not world domination, mind control or the ability to fly, but a life where you will never have to fight your siblings for food: 'above all they dream of food-pots that will cook endless porridge, tables or table-cloths that set themselves with meals.'[11]

There is an old French folk tale about a wonderful substance called 'fairy bread'. This was a food so abundant it could never be eaten up, with the proviso that it must never be shared with strangers. As Weber puts it, 'generosity stopped at the family threshold'. The fairy-tale dream – the happily ever after – is to reach a state of abundance great enough that no parent is tempted to make cruel selections among the many mouths at the dinner table. In a house where the bread never runs out, stepsisters might be nothing to be feared. They might become your friends. Then your only worries would be about the truly evil monsters – the witches or ogres, the ones outside the family who want to eat you up.

There is one category of people, on the other hand, who have never had to worry about siblings stealing the choicest morsels. The only child is freed from all those dinner-table squabbles. Those of us who were always in the shadow or shelter of another person often wonder if it would have been liberating to be the only small mouth at the table. How different life would be if you could blow out the candles on your birthday cake in peace and quiet, without some-one else breathing down your neck. Imagine never having to share your sweets. Chinese propaganda for the one-child policy when it was first introduced in the 1970s emphasized that singletons would have a larger share of resources: the nation's booty as well as that of the family. Freed from the 'meaningless uproar' of many siblings, these children would be healthier, better educated and better fed.[12] That was the theory, anyway.

The contrasting negative stereotype about only children is that they are spoiled and self-centred. The wealthy business-man George Hearst (1820–91) noted a certain selfishness in the eating habits of his son William Randolph, an only child who would go on to be one of the greatest newspapermen in the United States and the inspiration for the main character of *Citizen Kane*. 'There's one thing sure about my boy Bill. I've been watching him, and I notice that when he wants cake, he wants cake; and he wants it now. And I notice that after a while he gets his cake.'[13]

This behaviour probably had more to do with Hearst's personality than the fact that he was an only child. Several large-scale studies of only children – or 'singletons' – have disputed the common assumption that a lack of siblings makes people socially maladjusted. Research suggests that they are very similar in their behaviour and outlook to other children.[14] As for the idea that they are bad at sharing food, Lauren Sandler, the author of *One and Only*, a celebration of being an only child, makes the point that only children may actually learn to share food more nicely than others: they are copying the behaviour of adults, who have learned a bit of give and take, rather than modelling their behaviour on an immature sibling.[15]

Yet there are disadvantages to eating as an only child. In an affluent setting, having no siblings places you at a greater risk of childhood obesity. In one study, twice as many eleven-year-old 'onlies' were overweight as compared with children the same age with one or more sibling.[16] Why? The obvious explanation is that they are given all the goodies. One adult who loved being an only child recalled the joy of waking up on Easter Day and having twelve chocolate eggs and 'only me and mother and father to eat them'.[17] There are also signs that only children tend to get less exercise, without siblings to knock a ball around with.

But these are only tendencies, not rules. Depending on the family dynamic, almost all the health negatives of being 'only' could become positives. Maybe you exercise more, not less, because your parents have more time to take you to sports clubs or to the park with a Frisbee. It is possible that you develop *better* eating habits because your parents invest time each morning making you a healthy breakfast. Ultimately, each of us will never know whether we would have eaten differently with or without siblings.

Another question we will never know the answer to is how different our mealtimes might have been if we had been born a different sex. Before I was born, my parents thought I would be a boy, called Gabriel. Would he have given rise to a different repertoire of family dinners? Would he have agonized less over what he ate? When my sister and I were at the height of our disordered eating, I sometimes wished we had a younger brother, someone for whom food could be a more straightforward affair, instead of an endless psychodrama. He would eat hearty meals of bacon and eggs and never worry how the food was affecting his looks. I didn't realize then that boys get eating disorders too. Nor did I ever consider the possibility that a boy's presence at the table might have created more psychodrama over food for the sisters, rather than less.

It isn't only among impoverished families in rural India that brothers are encouraged to eat more while sisters are made to feel they should have less. It happens in richer societies too, the difference being that it is not classified as neglect but something that is good for the girls. Many girls pick up the message at home that their appetite is a problem, something that must be curtailed. We speak of 'growing boys', praising them for their manliness, but almost never of 'growing girls'. Maybe it is because we are fearful and embarrassed by the way girls' bodies grow – outwards as well as upwards. Is it any wonder that some teenage girls try to eat like their prepubescent sisters?

In most ways, parents learn to stop interfering in their

children's eating habits as they get older. Once children have their own money, we can no longer view ourselves as the controller-general of snacks. We become reacquainted with the pleasure of planning our own meals, without worrying that one child won't eat it if it has pepper in it while another will complain if it doesn't. In short, as our children grow up, we relax. The big exception is when parents think it is still their job to get a child, especially a girl, to lose weight, in which case the dietary pressure may continue and even intensify into adulthood.

I have sat at a table with grown-up children and their parents where a boy – in fact, a thirty-year-old man and not worryingly skinny – was told by his mother that he simply must have a second helping, because he needed it, while his sister was tutted at for accepting a single potato. In 1970s working-class France, as described by Pierre Bourdieu, such double standards were part of the culture.[18] The rule was abundance for males and restriction for females: meat eaten in great gulps for men; salad and tiny nibbles for women. A boy's path to manhood was marked by the privilege of being able to have your plate filled twice. A girl's journey to womanhood, on the other hand, was marked by self-denial. She must learn, like the other women, to share one portion between two and stay on her feet serving the men as they lolled in their chairs.

The original rationale for boys being encouraged to eat more than girls was the fact that men worked outside the home more than women, and needed plenty of food to perform their manual labour. Among the leisured upper classes of Europe, by contrast, it was sometimes seen as legitimate for girls rather than boys to overeat, because extra food would enhance their looks: think of the rounded white skin of the women in Veronese's art (c.1555–85) or the plump pink arms of girls painted by Boucher two centuries later. Now, plumpness is associated with poverty, given that energy-dense, carb-heavy food is generally much cheaper to buy than fresh healthy produce. In former centuries, however, for a woman to look 'well-fed' was a sign of wealth; and, as the twisted logic went, of beauty. The French gastronome and philosopher of food Jean Anthelme Brillat-Savarin, author of *The Physiology of Taste* (1825), felt sorry for women who were too thin, with the 'pallor of illness', and urged them to fatten themselves up on a rich diet with 'plenty of bread', hot chocolate, fresh eggs scrambled in butter, lots of meat, fish and soup with 'dishes made with rice or macaroni' and desserts of Savoy biscuits, babas and the like, plus well-sweetened fruits and grapes.[19] And beer.

But among the masses, beer belonged – and to a large extent, still belongs – to men, along with salty cheeses, salami and second helpings. To be male was to eat with freedom and a sense of largesse. 'It is part of men's status to eat and to eat well,' notes Bourdieu, though this does not explain why women were not thought to need sustenance for all the work they too did, serving, clearing, washing, cooking.[20] Women were supposed to be content with less, to have such 'dainty' tastes that they did not even aspire to drink strong liquor and eat strong meat, to derive their satisfaction from seeing their brothers and fathers well fed. A man with a small appetite, meanwhile, or one repulsed by meat, was viewed as 'suspect'.

From the 1980s onwards, as more equal patterns of work emerged, this rationale for sisters eating less than brothers dropped away, to be replaced by a new ideology of slimness, which again dictated that they eat less. 'If boys get fat they say it's muscle,' an eleven-year-old girl told researchers conducting a survey of children's eating habits in Europe, somewhat bitterly, in 1994.[21] Girls are still supposed to be content with less than their brothers, and to take up less space around the table. The difference is that parents who pressurize girls to slim down think they are doing them a favour, because they fear we live in a world where the consequences of being overweight are worse for girls than for boys. In this, they are not entirely wrong. A recent study in New York found a direct negative correlation between the extent to which women were overweight and their expected income, their ability to acquire a job and even to have a fulfilling family life.[22] None of these correlations held good for overweight men. Unfairly, life was significantly worse for the overweight women than for their male counterparts. Interestingly, the researchers could not decide if this was because society really does discriminate against overweight women or because they have lower self-esteem, which in turn leads to diffidence over applying for a job or a pay rise.

It wouldn't be surprising if they did have low self-esteem, given that girls – overweight or otherwise – are often made to feel that their eating is more of a problem than that of their brothers. Mothers who exert weight pressure may be projecting their own body anxieties onto their daughter, which explains why they do it more to girls than boys. (One study showed that bulimic mothers are more likely to exert 'controlling feeding practices' on daughters than sons, regardless of whether the girls are overweight.[23]) Fathers may have the same bias towards children of the same sex. Fathers who were dissatisfied with their body, one study found, were likely to monitor their sons' but not their daughters' food intake. But for the most part it is mothers who are in charge of the family's food and it is daughters who are given the lion's share of the weight pressure.

Parents, it seems, find it hard to know quite how to behave around teenagers at the table, whether they are boys or girls. Project EAT in Minnesota followed nearly 5,000 adolescents over a five-year period.[24] Of these 5,000, some were younger teenagers (aged twelve to seventeen during the time-frame of the study) and some were older (aged fifteen to twenty). Researchers led by Katherine Bauer used interviews with the adolescents to assess any changes over the years in how much parents encouraged them to choose healthy foods, to exercise and to lose weight. Over time, parents of boys backed off on all three counts. This reflects the growing independence of the teenage years. It feels slightly absurd to be asking someone bigger, stronger and hairier than you if they have eaten five vegetables yet today. Maybe there's also a shift of priorities. When you fear that your teenage boy might be into drugs, it can be hard to care so much about the odd fizzy drink. Parents of teenage boys don't want to be too heavy about family meals, for fear of driving them away altogether, which is a pity, because the data suggests that continuing to give them balanced home-cooked meals is one of the best things you can do for your adolescent's health. Children who eat regular home-cooked meals consume more fruit and vegetables and seem to be happier, insofar as this can be measured.[25]

With teenage girls, however, the picture was different. As with the boys, the girls reported that as they got older, their parents made less strenuous efforts to get them to eat healthy food and to be physically active. But as the older group moved from mid to late adolescence – from fifteen to twenty – some of the parents actually started to put *more* pressure on them to go on weight-loss diets. At just the stage of life that we'd expect a girl to be given full independence in matters of food – at an age when they were free to drive their own car, get married, vote – rising numbers of them were being told by their parents that what they ate was still the family's business.

Our childhood experiences with food can trap us in destructive patterns for the rest of our lives. This pressure on the girls in some families to lose weight is a case in point. I recently spoke to a professional woman in her late forties who had been on more failed diets than she cared to remember. Every time she phoned home, no matter what else was happening in her life with her children or her career, her mother's first question was still, 'Have you lost weight?' Her friends cheerily joked that she would only stop the yo-yo dieting once and for all when her mother died. She felt that she had trapped herself in overeating for many years as a form of rebellion against her upbringing, but one that ended up as a form of self-punishment. At last, in her forties, she had found a form of healthy eating that she could maintain – not the diet that her mother wanted her to go on but a routine of lovely salads, grilled fish and intensely flavoured soups that did not feel like deprivation. It had taken her such a long time getting there and she felt she could have arrived sooner without the parental voice in her head telling her to eat less.

Like so many of the things parents do out of loving solicitude, putting pressure on girls to lose weight has no good outcomes and a lot of bad ones. It has repeatedly been shown to leave children more at risk of body dissatisfaction, excessive worry about weight, depression, bingeing and disordered eating. In one study, nearly a third of fifteen-year-old girls whose mothers encouraged them to lose weight engaged in extreme weight-control behaviours – including laxatives, vomiting, skipping meals, smoking as a meal substitute and diet pills – as against just 5 per cent of girls whose mothers did not encourage them to diet.[26] There are similar consequences from family 'weight teasing' of overweight girls. One study found that girls who were teased 'very often' for being chubby by relatives were far more likely to resort to binge eating than those with more courteous families.[27] Even if this doesn't deter you from hinting to your daughter to slim down, consider this: it probably won't work. Having parents pressure an overweight child to diet left them with a higher risk of being overweight after five years.[28]

This weight pressure on girls is part of a wider mindset. The sphere of eating can feel like a different place for boys and girls. Regardless of whether we have actual brothers or actual sisters, our culture sends strong hints that we should choose foods appropriate to our sex. 'For a boy certain things were off-limits,' recounts food writer Nigel Slater in his memoir *Toast*. As an eight-year-old boy, Slater felt labelled as effeminate by his choice of treats. 'Love Hearts and Fab ice creams were for girls . . . and no one over six would be seen dead with a flying saucer.'[29] There's still a macho note to much of the food advertising aimed at boys and men. 'Man up!' urged KFC's slogan for its Double Decker Burger, as if there might be something wimpish in a boy who didn't think he could manage a bun containing two full chicken burgers, bacon strips, cheese, lettuce and a splodge of mayonnaise.

Girls, too, are given strong messages that some foods are more for them than others. When businesswomen in Japan socialize with the boss, they are 'allowed to drink plum wine mixed with plenty of soda instead of beer' reports the

Economist, as if it were impossible for a woman to prefer beer to watered-down plum wine.[30] Such thoughts go all the way back to the playground, where boys are 'meant' to want foot-long sandwiches stuffed with meat, while girls are supposed to dream of sugar and spice and all things nice. Before they are into double figures, girls learn that certain foods come with a 'shouldn't' attached. They learn that 'skinny' is a compliment. They absorb the sense that there would be something slightly odd about a girl who doesn't revere chocolate.

The idea that boys and girls respond to food differently is not all myth: there are definite physiological differences in the way that boys and girls relate to food. For one thing, there's the vexing fact that – levels of exercise and size being roughly equal – boys need more energy than girls. According to recent guidelines, a seven-year-old boy needs around 100 calories a day more than a girl of the same age (1,630 versus 1,530). By the age of eighteen, the gap jumps to nearly 700 calories (3,155 calories for boys versus 2,462 for girls), the equivalent of a whole extra meal.[31] There's a difference between pressuring a girl to eat less and simply giving her a different portion. If I fed the same mountainous helping to my average-height twelve-year-old girl as I give to my sixteen-year-old boy (6 foot 10), she wouldn't be able to make a dent in it. That isn't meanness; it's maths.

A more startling physiological difference is that male and female brains show different activity in response to eating. The sour taste of citric acid, for instance, has been found to produce a more marked response in the insula and thalamus of women than men. Overall, women have a greater sensitivity to smells and flavours than men and are better at remembering them.[32] This increased sensitivity may make girls pickier. Many studies have shown that women have a greater number of negative attitudes towards foods and are more likely to reject them for not tasting quite right. Marketing expert Bryan Urbick has spent many years working with panels of children doing product development for the food industry, interviewing on average 4,000 children a year in Europe, North America, Central America, Asia and the Middle East. Across these different cultures, Urbick found 'strong, recurring patterns' in the ways that girls and boys responded to food. Boys regularly had lower sensitivity to taste and touch than girls. Urbick found that 'if you get the taste right for girls, you have probably got it right for boys as well'. Yet when it came to the branding of food products, the situation was reversed. Urbick advises product developers that when targeting both boys and girls, the packaging and brand should always be skewed to boys. 'Girls are more accepting of "boy" products whereas boys are more likely to reject something that is too "girly".'[33]

The idea that there are 'girly' foods and 'boyish' foods is pretty ridiculous when you think about it. Who gets to decide that *Men Love Pies, Girls Like Hummus*, the title of a 2013 cookbook by chef Simon Rimmer? Yet even if we insist we are beyond such childish notions, it is very hard not to internalize the idea that some foods belong more to one sex than others and to make our choices accordingly. We tend to automatically associate hearty meat dishes with men and lighter salads and sweets with women, and these stereotypes are replicated in cultures as different as France and Japan. When American college students were asked to say which foods belonged to which sex, they answered readily: steak, fried potatoes, onions and hardtack candy were for men and cottage cheese, peaches, soufflé and crêpes were womanly.[34] Moreover, it was found that some adolescent boys were wary of eating foods with feminine connotations, above all when friends were present: real men don't eat soufflé!

These ideas about boy food and girl food tip over inexorably into our personal preferences. In 2003 one research group used survey data to explore how the concept of 'comfort food' differed according to gender.[35] They found that 'males preferred warm, hearty, meal-related comfort foods (such as steak, casseroles and soup) while females preferred comfort foods that were more snack related (chocolate and ice cream)'. Striking numbers of women also considered vegetables to be a comfort food. The researchers suggested that the fact that men were comforted by full hot meals was likely to be a throwback to childhood, when they became 'accustomed to having meals prepared for them'. Another gender difference was that men tended to feel 'healthy' after eating their preferred comfort food of steak whereas women felt 'guilty' after their ice cream, biscuits and chocolate. So the women's comfort food didn't even comfort them: a waste all round.

This social pressure to eat according to your sex matters more than it seems. For one thing, it undermines our pleasure in eating, which is seldom a good idea. Women often deny themselves the dish they might really want on the menu because they feel it isn't 'appropriate'.[36] Women in Japan reported intense cravings for sushi, but did not eat it as often as they wished, perhaps because it is seen as a filling, masculine sort of food.[37] A survey of how British consumers behave when eating out found that, while women may say that they enjoy steak in restaurants, they do not ultimately choose it as often as men, seeing it as too expensive for them.[38] Overwhelmingly, female diners-out choose white meat and male diners choose red meat. We see nothing wrong in this because we are so acclimatized to the view that men are the 'red-blooded' ones when it comes to food.

But if anyone needs red meat (and some would argue that no one does) it is not men but adolescent girls. What is so damaging about our gendered approach to food is that it encourages both boys and girls to feed themselves in ways that go against what their bodies require. We have got things the wrong way round. It is girls more than boys who need the most haemoglobin-boosting foods. And boys more than girls are lacking in salad and vegetables. Girl food and boy food are dangerous nonsense that prevent us from seeing the real problems of feeding boys and girls.

Contrary to the impression given by the health pages of newspapers, the greatest single nutritional shortfall in our diets right now is not our failure to eat enough 'superfoods', whatever those might be. It is the iron deficiency of girls. Across the globe, rich or poor, fat or thin, millions of adolescent girls are anaemic because they do not have enough iron-rich foods in their diet to cover the leap in what their bodies need – from 8mg to 15mg – when they start to menstruate. Many more have iron depletion, where no iron is stored in the body, which can cause tiredness, headaches and impaired cognitive function.[39] Iron deficiency affects 2 billion people worldwide, according to the World Health Organization. Plenty of men and boys are anaemic too, but young women are disproportionately affected. It takes its heaviest toll in developing countries, causing one in five deaths in childbirth (often due to haemorrhage; a pregnant woman needs twice as much iron again as a teenage girl). But it is also remarkably common in countries where you might expect girls to be well nourished, and isn't helped by the stereotype that girls can survive on lettuce leaves and chocolate.

A European survey from 2001 found that as many as 40 per cent of girls aged fifteen to sixteen in Sweden had depleted iron stores (as against 15 per cent for boys); in Denmark it affected 7 per cent of boys aged sixteen to seventeen and 20 per cent of girls of the same age.[40] In China a sample of 1,037 adolescent girls in 2007 found iron deficiency in 40.4 per cent and full anaemia in 19.5 per cent.[41] Short of supplements (which can cause constipation and nausea) and fortified breakfast cereals, it isn't easy to eat enough iron-rich foods. By far the richest and most 'bioavailable' source is liver (a 3oz serving of chicken liver contains 11mg) followed by red meats (a 6oz sirloin steak: around 6mg iron).[42] Absorption of iron is impaired when tea or coffee is drunk alongside, whereas vitamin C improves absorption.

A teenage girl who suddenly switches her morning juice for coffee, turns vegetarian and stops eating eggs for breakfast (eggs are boy food!) may find herself in trouble. Women are far more likely to be vegan or vegetarian than men: out of around a million American vegans, 79 per cent are women; for vegetarians, 59 per cent are female. There are plenty of good vegetarian sources of iron including pumpkin seeds, nuts, spices, green leafy vegetables, blackstrap molasses, dried fruits, egg yolks, pulses such as kidney beans, bran and wholemeal bread. The problem is that a child who turns vegetarian on impulse is unlikely to be in the habit of eating the correct balance of these foods. Non-meat sources of iron are also less readily absorbed by the body, so vegetarians may need even more iron than their carnivorous counterparts. The most dangerous group for girls to be in for anaemia are those who are both dieters and vegetarians: in the 1990s 43 per cent of vegetarian British girls aged eleven to fourteen who had tried to lose weight over the past year had low haemoglobin levels in their blood, whereas only 15 per cent of non-dieting vegetarians did, and 8 per cent of non-dieting meat eaters.[43]

Surprisingly, the overweight are also at high risk for anaemia. The problem is due not to the amount of food they consume, but the quality. A sample of Iranian adolescent girls found that despite taking in more calories than they needed overweight girls were actually more likely to be anaemic than others: 34.1 per cent as against 27.8 per cent.[44] These girls are not getting enough iron from their carb-heavy, nutritionally imbalanced meals. There's not much iron in pizza, or ice cream or chips, all newly popular foods on the streets of Tehran.

It may be that iron deficiency is cause as well as consequence of being overweight. Iron deficiency seems to slow down metabolism, through a reduction in carnitine levels (a compound involved in metabolizing fatty acids).[45] Anaemic women have a poor capacity for aerobic exercise, which can be reversed with iron supplements. It's understandable if teenagers with anaemia are not raring to get on a treadmill, given that lack of iron makes you feel faint and woozy, hardly able to put one foot in front of the other (or so I felt when I was anaemic and as white as a sheet after the birth of my first child).

Too many overweight teenage girls are being pressurized by their families to lose weight when what they in fact need is better nourishment (which would almost certainly help them to lose weight along the way). Iron-deprived girls of every size need to be 'built up', as the reassuring old phrase had it, with soft-boiled eggs and wholemeal toast soldiers; dark leafy greens; griddled flank steak; lamb casserole; grilled sardines; and hearty minestrone or black bean chilli. In short, these girls need man food. If they still feel their body is crying out for chocolate, they should switch from cheap milk chocolate bars which are really mostly sugar and vegetable fat to real 70 per cent cocoa dark chocolate which might actually do them some good (a small 30g bar of dark chocolate: 5mg of iron).

But almost no one speaks of building up girls – these delicate soufflé-nibbling creatures who ought to be able to survive on air and compliments. Instead, great focus is placed on building up their brothers, stoking them with dangerous delusions about how much food they need. In the current food environment, the overfeeding of boys is no more helpful than the underfeeding of girls.

One of the biggest obstacles to any overweight or obese child losing weight is when parents do not see that there is anything wrong. Dr Laura Stewart runs child obesity clinics in Tayside in Scotland, which has some of the worst child health outcomes in Europe. 'A lot of parents do not perceive their child has weight problems,' Stewart told a conference of health professionals I attended in October 2013.[46] The problem, in Stewart's view, is that the media only shows photographs of 'the most exceedingly obese children' when discussing the issue, which leaves people whose children are only mildly obese thinking that they are perfectly fine. If most of the people you see around you are overweight, it starts to seem normal.

Stewart told us of an experiment she did in Tayside in 2010, which was aimed at finding out what it would take to make parents see that their child had a weight problem. Parents were shown photographs of children and asked to group them into 'normal' and 'overweight'. It was as if they couldn't grasp what was in front of them. These parents could only recognize that a child was overweight when they were actually 'significantly obese'. All the photos of children who were overweight rather than obese were seen as 'normal'. There is a lot of collective denial when it comes to children and weight. It is even there among some of the doctors and nurses whose job it is to deal with it. Around half of those employed by the NHS are now overweight or obese, making it seem normal for them too. After Stewart had finished talking, a health worker in the audience put up her hand and criticized Stewart for using the BMI system to measure children. 'You could have a fat but fit child! On the BMI system, a body-builder could be classified as obese!' Stewart collected herself and smiled. 'I've never once had a fit muscular child referred to my clinic. You do have to use your eyes.'

The problem of 'weight misperception' has consistently been shown to be worse with boys than girls. Numerous studies have shown that parents are less likely to think a brother is overweight than a sister, perhaps because a boy's figure is seen as less of a social issue than a girl's. Parents often become very defensive when a nurse or dietician points out that based on the weight and height charts, their child is obese. They protest that their child is 'big-boned', that they only eat healthy food, that there's something wrong with the chart. This defensiveness is seen in parents of girls as well as boys, but with boys, the parents seem almost proud of their son's large physique. An obese boy may be described as 'solid' or 'chunky' which sounds healthier than 'fat'. A low-income mother whose preschool son was borderline obese told researchers, 'I can look at him, and he might weigh a lot, of course, but that could be just all muscle. 'Cause he is a strong kid.'[47]

When parents under-assess a boy's weight and over-assess his food requirements, it may form his self-image and the way he eats for the rest of his life. A large-scale study of weight perception among more than 16,000 American adults found that under-assessment of one's weight was two to three times more common in men than women.[48] Nearly 43 per cent of men who were overweight (but not obese) said that their weight was 'about right'. Nearly 12 per cent of obese men said their weight was about right. Similar results have been seen among Australian men and women, where about the most men would admit to was that they were 'a little over-weight' when they were actually 'biomedically obese'.[49] This is worrying, given that you are unlikely to do anything about your weight problem if you don't even recognize that you have one.

There is an equally disquieting trend for a large segment of healthy-weight or underweight women to misdiagnose their weight in the opposite direction and wrongly label themselves as obese. A 2003 study of more than 2,000 undergraduates across six American campuses found that fully 72 per cent of women said that their 'thighs were too fat' in comparison with just 11 per cent of men.[50] It transpires that, regardless of actual weight, college-age women are more uncomfortable about being weighed in public than their male equivalents.[51] Female body dysmorphia – exaggerated and perfectionist beliefs about one's own body flaws – is an anxiety that is not confined to those who suffer from a full-blown eating disorder. Research suggests that in our society such worries afflict the majority of post-pubescent girls.[52]

This discrepancy between the self-image of girls and boys is a problem for both sexes. It means that most public health messages about the 'obesity crisis' and ways to eat more healthily are probably heard by the wrong people and not heard by those who might benefit. It's like when a teacher loses their temper with the persistent troublemakers and decides the best approach would be to tell off the whole class to make them see how they are squandering their future. The good quiet children listen to the angry lecture and feel upset, thinking it applies to them, when it doesn't. The troublemakers at the back don't think the teacher has anything of relevance to say to them, so they doze off or continue to make trouble. When anti-obesity campaigns tell us we are sleepwalking into a terrible future, significant numbers of healthy-weight women think the words are a direct criticism of their own upper arms and significant numbers of obese men think it has nothing to do with them.

As a general rule, women seem to be more compliant with dietary guidelines, more likely at least to attempt to eat healthily even if they do not manage it.[53] Over the course of childhood, the trend is for boys to eat declining quantities of vegetables and fruit, while girls increase their intake very slightly. A recent study of British children showed that 70 per cent of boys aged four to six ate apples but this declined to 39 per cent of those aged fifteen to eighteen; as we have seen, parents feel reluctant to push older boys on the question of healthy eating.[54] But with girls, the consumption of salad went up: only half of four- to six-year-olds ate it, but 66 per cent of teenage girls did, maybe because they saw it as something they were meant to eat. Either way, many boys could benefit from learning to eat more like a girl. In Thailand, girls eat much more fruit and vegetables than boys, which is reflected in obesity rates: twice as many boys are obese.[55] Yet survey data suggests that Thai mothers do not view their sons' lack of vegetables as a problem.

Because we don't recognize obesity in boys to be as much of an issue as it is for girls, we also don't see just how unhappy it can make them. Kuwait has some of the highest rates of adolescent obesity anywhere, with nearly half of those aged fourteen to nineteen overweight or obese. Kuwaiti teenagers are significantly fatter than those in other Arab countries. One study found that more than twice as many Kuwaiti boys aged fifteen to eighteen were obese or overweight, as against those in Syria or Libya.[56] Kuwait had a more rapid 'nutrition transition' than other parts of the Middle East, meaning Kuwaitis were exceptionally quick to adopt Western fast food. For generations, Kuwait has eaten a cuisine far heavier in meat and fat than nearby Arab countries such as Lebanon, where family meals centre on vegetable-based mezze: aubergine dip, fattoush (herb and bread salad) and the like. In Kuwait, a typical family dish is Makbous Dajaj, where a whole chicken is first boiled and then deep-fried and served with rice simmered in the fatty cooking water.[57] When American fried chicken and hamburgers arrived in Kuwait City, local palates were already primed to enjoy them.

Kuwait's obesity crisis is another case of traditional ideas about feeding going badly wrong in a modern prosperous food supply. Hospitality is a crucial concept in Kuwait, and it is sometimes remarked that every family gathering is treated as if it were Thanksgiving. Claudia Roden, the great expert on Middle Eastern food, notes that to be a good guest in traditional Arab cultures, 'if one feels satiated, one should nevertheless continue to nibble at a dish from which others are eating, since if one person stops eating, everyone else may feel compelled to stop too'.[58] Oil money has made Kuwait one of the richest countries in the world per capita, with vast shopping malls, a culture of eating out and luxury cars, and plenty of disposable cash for snacking. The wealth of Kuwait means that excess food has become affordable to many, hence the fact that weight gain has been far more rapid than in poorer countries such as Syria or Algeria.

Kuwait has also suffered an epidemic of 'disordered eating', particularly for boys. Professor Abdulrahman O. Musaiger, a leading expert on nutrition in the Arab world, found that eating disorders were becoming widespread throughout the region and the teenagers with the most disordered attitude to eating of all were Kuwaiti boys.[59] Musaiger tested for eating disorders using the EAT-26 test which asks you to rate the frequency with which you engage in behaviour such as 'eating binges where you feel you may not be able to stop' and thoughts such as 'I feel that food controls my life' or 'I feel extremely guilty after eating'. A full 47 per cent of the Kuwaiti teenage boys who performed the test came out as having 'disordered eating attitudes' (a lot of Kuwaiti girls had disordered eating too: 43 per cent). Musaiger concludes that this might be because Kuwaiti boys have 'cultural adjustment difficulties'. Despite its economic modernity, the country is socially more conservative than many other Middle Eastern countries. As recently as 2013 a survey of nearly 2,000 college students found that only 70 per cent of these young Kuwaitis – male or female – believed that men and women were equal.[60] The survey also found that many students claimed to dislike the Western 'consumer culture'. Kuwaiti boys hear relatives speak of the Gulf crisis of 1990 and may view America as the enemy.

Yet even as they denounce American culture, they participate in it when they grab a Super Supreme at Pizza Hut (which has forty-nine branches in Kuwait) or a platter of spicy ribs at Applebee's. Many Kuwaiti boys eat the enemy's food and sip the enemy's fizzy drinks every day. In Professor Musaiger's analysis, these young men are torn between the values of East and West, 'caught between the influence of Western culture on dietary behavior and body size preferences that encourage thinness on the one hand and traditional cultural norms that favor customary food habits and normal or plump body size on the other'.[61] Traditionally, it is a boy's prerogative to eat as much as he wishes. In a Kuwaiti family where all the brothers are obese – and most of the adults around them too – the parents may look at their boys and think they are fine. But judging from Musaiger's research, they are absolutely not fine, either physically or mentally.

There is a great deal more to Kuwait's obesity crisis than just gender. But wherever we live, one step in learning to eat better would be to move beyond gendered notions of food. Both boys and girls would be better off if they could copy the best aspects of each other's eating habits. Boys could do with being more conscientious about vegetables and more honest about their body size. Girls would benefit from adopting a manly liking for hearty main courses instead of sugary pink cupcakes and chocolate. Like boys, they should know that they have permission to eat when they are hungry.

The great potential of siblings – or the virtual siblings we assemble among our friends – is that they make us less alone at the table. We borrow tastes and habits from them. Maybe their example makes us see that one tiny bowl of cornflakes is not a big enough breakfast; or, conversely, that a crunchy stick of celery can be surprisingly delicious, especially if you add peanut butter and raisins. When they go through a craze for Moroccan takeaways, or wok cooking, or growing herbs, it broadens our horizons too. And – when we are not squabbling – their company makes everything taste better.

Siblings don't have to be rivals at the table. These days, my sister and I live on different continents, so we don't eat together as often as I'd like. (She bagged America; I was left with Europe.) But when we do meet, the dynamic of eating is quite different from what it once was. Now that we are middle-aged and have our own children, we have both calmed down and the result is the sort of culinary convergence I would once never have thought possible. It turns out that we are not so different when we sit down to eat side by side. We both like: very strong coffee; toasted bagels with butter not cream cheese; avocado sushi rolls; and any kind of fruit, especially crisp tart apples in the autumn or ripe juicy pears, which remind us of our mother. When I'm visiting my sister, we sometimes go to a Vietnamese deli near her house that makes delicious sandwiches. She orders a tofu banh mi: soy-marinated tofu with crisp coriander leaves, pickled carrots and mooli radish in a soft hoagie roll. I pause for a moment, consider going for roast beef on rye just to prove I am my own person; and then I choose exactly the same.

CHOCOLATE

'For some reason,' writes marketing expert Bryan Urbick, 'girls have a special relationship with chocolate.'[62] Almost all the ads for chocolate on TV are aimed at women, depicting them as powerless to resist its melting charms.

Chocolate's status as food for girls and women – something to be yearned for and then regretted – is so ingrained that it would be easy to assume that there is something deep within females that makes them crave the chemicals in chocolate. Girls often talk this way themselves, saying that they 'need' the happy serotonin of chocolate because they have their period.

There are, no question, some potent chemicals in chocolate. These include phenylethylamine (an amphetamine-like substance), caffeine and anandamide (a cannabinoid). Yet the form in which most women crave chocolate is milk rather than dark chocolate, which is significantly lower in these compounds and higher in sugar and fat. If a 'chocoholic' is craving anything, it is probably the dopamine release of sugar. The idea that menstruation hormones make women crave chocolate has been undermined by a recent study[63] in which researchers found that post-menopausal women only experienced a very slight drop in their chocolate cravings, despite the fact that they no longer had periods. The researchers concluded that stress, not hormones, drove women to chocolate.

Women have a special relationship with chocolate mostly because our culture tells them to. It goes back to the old claptrap about sweet treats being for 'ladies', while savoury tastes are for men. Chocolate is undoubtedly an appealing substance: the heady aroma, the sweet taste, the way it melts at body temperature. But there's no biological imperative that says that women should be driven to seek these experiences more than men.

In 2006 a fascinating study was done on students in Spain and the US.[64] Among the Americans only 59 per cent of men would admit to chocolate cravings as against 91 per cent of women. In Spain, on the other hand, chocolate cravings were much more evenly split: 78 per cent of Spanish men and 90 per cent of Spanish women said they craved it. This is a clear indication that the female craving for chocolate is something that is culturally determined, not innate.

Female chocolate cravings are an archetypal learned behaviour. From our earliest years, girls pick up on the fact that chocolate is special and *for us*. We are given it at birthdays and holidays or to calm our tears. We absorb the message that chocolate will soothe us when we are down; and that when we are happy, it will make us happier still. We tell ourselves we don't just want it but need it. Yet buying it also seems to leave more women than men feeling guilty.[65]

A girl's physiology does not make her need slabs of chocolate once a month, any more than it creates a requirement for marshmallows, macaroons and salted caramel eclairs. If our chocolate habit is learned then – unlikely as it seems – it can be unlearned, or at least toned down a touch.

CHAPTER 6:
Hunger

So it happens that when I write of hunger,
I am really writing about love . . .

           M.F.K. FISHER, *The Art of Eating*

A child feels woozy, slips under his desk and onto the floor. Another child yawns for hours every morning, only properly waking up after lunch. A third seems incapable of concentrating on even the simplest lesson, as if his brain isn't quite there.

Such hunger wasn't meant to exist any more in civilized societies where food is plentiful, but people who work for hunger charities attest that in some of the schools of the affluent West scenes like this are a daily occurrence. The No Kid Hungry initiative works at a state and city level to help provide free breakfasts to low-income children in 25,000 schools across America, on the premise that a child who has not eaten in the morning is not ready for learning.[1] Giving a child a bowl of cereal, some fruit and a carton of milk is a small thing. On the other hand, viewed from the perspective of an individual child's future, feeling full in the morning is not small. It can make the difference between being switched on to education and all the advantages this brings; or not. People who work for No Kid Hungry have found that compared to other low-income pupils, the children who eat a free school breakfast have higher maths scores, better attendance rates, are more likely to graduate from high school and, perhaps most critically, are less likely to experience hunger as adults. Once the pattern of not being hungry in the morning is set, it modifies the way you eat for life.

Satisfying hunger is the most basic function of eating. Assuming there's enough food – and this cannot always be assumed, as the presence of food banks and school breakfast programmes reminds us – hunger management doesn't seem like something anyone should have to learn. Unlike our appetite for specific foods, hunger is an innate animal mechanism that we are born with. Yet as it plays out in the modern food environment, hunger is far from simple. Beyond infancy, acquiring the ability to cancel out our hunger adequately without overshooting the mark has become a complicated task, whether in the hungry developing world or the over-stuffed West.

Like boredom, moderate hunger is one of those childhood discomforts that modern parenting plus a world of abundance seems to have all but abolished in middle-class families. Children and adults alike are constantly topped up with snacks to keep bad moods at bay. My handbag has a compartment filled with cashews, cereal bars and dried fruit in readiness for those frequent moments when they groan, 'We're huuuuuungry', these children of mine who have never known the gnawing emptiness of real hunger. Hunger now tends to be nipped in the bud before the first growl of the tummy. Yet when it comes to bona fide child malnutrition, we are still far too willing to pretend it doesn't exist. While malnutrition in poor countries remains the number one cause of child mortality in the world, its effects can also be seen closer to home.

The rise of child obesity creates the illusion that our problems with eating are all about overfeeding and we find it hard to recognize malnutrition even when it is right in front of us. Some are sceptical that any child in the developed world is truly hungry. Yet Feeding America, the largest hunger relief programme in the US, estimates that 15.8 million American children live in 'food insecure' households and therefore grapple with hunger, at least some of the time. The median household income of the families using Feeding America's food banks was just $9,175, as of 2014.[2] Eighty-four per cent of the families they see report buying the cheapest food available, regardless of whether it was healthy. Cheap food doesn't have to be food that is unwholesome, unsatisfying and lacking in essential nutrients; but, as it happens, it mostly is. Many children today in affluent countries are in the paradoxical situation of being overweight and malnourished at the same time (this is known as 'hidden hunger'), taking in too many calories from cheap carbohydrates without enough of the micronutrients the body requires. Plenty of others, however, are hungry in the old-fashioned sense, light-headed from a lack of food, particularly from lack of the protein foods children need to grow. And there is no hunger so ravenous as that of a child.

Teachers report children fainting in the playground; or those who come to school wired on a breakfast of sweets and fizzy drinks or leftover scraps of fried takeaway; and in rare cases, boys and girls whose stomachs are distended, like the African children on the TV news. The charity Magic Breakfast was founded by former businesswoman Carmel McConnell in 2001 when she visited inner-city London schools as research for a book and was shocked to hear of pupils complaining of hunger. Now, Magic Breakfast supplies more than 8,000 children with healthy breakfasts through morning 'clubs': cold cereals and milk, fresh fruit and special protein-enriched bagels served with butter and jam, and sometimes porridge, which fills the children up if only they can be persuaded to eat it.

On a bright July morning, I eat a Magic Breakfast at Keyworth Primary, an 'outstanding' school in a deprived area of south London. The children – aged four to eleven – are eating buttered bagels and sliced-up wedges of apples and oranges. There's a good toasty smell in the room. The atmosphere is calm, with a group of soccer-mad boys at one end of the table and a gaggle of girls chattering away at the other end. The school's head teacher, Susi Whittome, remembers that before the breakfast club was introduced, children often arrived at school either sleepy or 'upset', with many kids going wild or zoning out in lessons, which turned out to be a response to hunger. Both behaviour and educational results at the school have greatly improved now that the children are not hungry. Whittome wants children to leave her school as 'ambassadors for social justice'; not easily done on an empty stomach.

The symptoms of long-term hunger are horrible and complex, all the more so in a child. Apart from loss of weight, the child may feel tired and cold all the time, and worried, depressed or disengaged. They bruise easily and lack focus. Their skin is dull and they may have permanent goosebumps and cracked lips. 'White lips on black children' was one of the signs of hunger observed by Marvin Davies, a social worker and teacher working with agricultural labourers and their families in Florida in the 1960s.[3] Prolonged chronic malnutrition affects children for the rest of their lives because it affects both brain development and growth. One in four children in the world are stunted by hunger and many of the consequences are irreversible.[4] Yet in the short term, food has a power to reverse the symptoms of malnutrition very quickly. After just three or four weeks of adequate feeding, a child's body may get back in sync: anti-hunger charities in the developing world cite similarly quick rates of recovery from feeding interventions. Magic Breakfast say they see children perked up by the very first bagel, which may give them enough energy to stay awake for that morning's lessons. The psychological consequences of hunger, needless to say, last longer.

When I sit and eat with the children at Keyworth Primary and ask them what they like about the breakfast, I expect them to talk mainly about the food. Certainly they are chewing their bagels with gusto. But what most of the children bring up is the social aspect. 'It's fun,' says one. 'The best bit is playing games,' says another; when the children have finished eating and cleared their dishes, they are allowed to catch up with homework or play board games like Monopoly and Connect Four. Three sisters are sitting in a row, laughing.

Many of these children live in housing so small that there is no room to eat together round a table. When I ask one of the girls what she eats for breakfast at weekends she says, 'Anything random.' It isn't just food that is lacking in their lives: it's the space to share it in and the mood of relaxation and security that comes from a structured communal meal. In 2011 a study of more than 22,000 low-income families with young children found that those who lived in crowded housing were more likely to suffer from food insecurity too.[5]

Likewise, a hundred years ago in Glasgow, investigators found that there was a direct correlation between the housing a child lived in and the extent to which they were underfed. Whereas a child from a house with four rooms weighed on average 64.9lb, and one who lived in three rooms averaged 60lb, a child from a one-room dwelling weighed just 52lb.[6] It may be that the size of home and the child's weight had a common cause in poverty: those families who could only afford one room were also those with the least money to spare for food. Either way, cramped accommodation is still one of the biggest obstacles to feeding a family well, because whether a meal is satisfying depends on the setting and company as well as the food.

A breakfast club gives these children something their hard-pressed families couldn't provide them with no matter how hard they tried: the chance to sit at a spacious table and eat with friends. Food – the right food – is clearly essential to preventing hunger. High-protein bagels are good, especially when the alternative is a growling tummy and a fuzzy head. But talking to these children at Keyworth Primary, it's clear that the craving that they needed to satisfy was for something more than just toast. What makes breakfast clubs such as this one so effective is that they feed the hunger for social interaction as well as the hunger for food.

Hunger runs deep. Before you can fully cancel it out, you need to learn what it is you are truly hungry for.

Being able to regulate the amount of food we eat according to our needs is perhaps the single most important skill when it comes to eating; and the one that we least often master. In contrast to its close relation, appetite, hunger looks like a basic impulse: an expression of the body's biological need for food. Yet the more you examine it, the more you see that hunger is not a simple drive at all. Hunger is always a kind of emptiness – an absence of nourishment – but what it will take to replenish it is far from obvious. Back when we were children, we learned to respond to hunger in ways that were only partly about our body's need for food. Maybe we suppressed our hunger so that we could eat less and lose weight, or we feigned it so we could eat more. Or we ignored it because we would rather carry on playing. Very little of the eating that happens in the modern world is as simple as: feel hungry, eat food. The great challenge for most people is learning how to recognize when we have had enough.

'The cry of the newborn for food,' says physiologist Anton J. Carlson, is hunger in its 'purest form', a pain that is immediately quieted with feeding.[7] For the rest of our lives, hunger will never be quite so pure again. With animals, the main test for hunger is eagerness for food, but it's one of our quirks as humans that we sometimes remain eager to eat even when our bellies are full to bursting. The existence of anorexia proves that it is possible to be hungry and yet not eager for food. Children will declare that they are stuffed, couldn't eat another morsel, pushing their main course aside, only to discover that they are, in fact, 'starving' when a tray of cupcakes appears. Many are quick to learn that 'hunger' is the ultimate justification for eating. As a teenager, I remember often claiming I was peckish when really I was just lonely and bored. It's far harder to refuse a child's request for snacks when they plead an empty stomach. Maybe grown-ups are not so different. I've lost track of the number of times I've heard waiters – hoping to sell something from the dessert menu – claim with an endearing smile that we have two separate stomachs, one for savoury, one for sweet.

Even discounting the little lies we tell ourselves to justify a slice of cheesecake, hunger is not easy to measure or define. The feeling of nausea or stomach ache – which ought to be the exact opposite of hunger – is remarkably similar to it in the early stages: rumbling in the gastrointestinal tract, stabbing pain in the abdomen, a restless feeling that your body is out of kilter and needs to be given something to fix it. Equally, it is possible to be so hungry that you do not feel hunger any more. Aid workers helping acutely malnourished children perform an 'appetite test'. If a child is so weakened that they refuse a small amount of food offered by a parent, they are in imminent danger and must be admitted to hospital. Among nutritionists, there is no universally recognized standard for measuring the sensation of hunger.[8] Common sense might say that hunger could be gauged by the amount of time since you last ate. If food is a fuel, we should feel that the tank is emptiest when the longest time has elapsed since it was filled – usually, in the morning. Unless you dabble in midnight feasts, the longest gap between meals is the overnight stretch before breakfast. Yet numerous trials have shown that people do not tend to be hungriest in the morning, despite the gap.[9]

Most people do not feel their peak hunger at weekday breakfast time (they are more likely to be hungry for weekend breakfast, another sign that hunger is a social drive).[10] Some – mystifyingly to me – have to be coaxed to nibble anything much in the morning. One of the great regrets of my life was pretending to be such a person when I was seventeen and on a language exchange with a family in France. On the first morning, they asked what I liked for breakfast. In my stilted French, I said I preferred just to drink black coffee, thinking this would make me look sophisticated (and yes, probably kidding myself that I'd come home as skinny as a

French girl). Day after day, I watched in silent envy as the family devoured crusty baguettes with fresh white butter and apricot jam with bowls of milky coffee and hot chocolate. I sipped my acrid black coffee. My pride would not allow me to admit how famished I was.

Apart from the gap between meals, another way to measure hunger is through various hormonal 'biomarkers'. These seem to promise a more objective, scientific way to determine whether a person is hungry or not, though in practice they confirm that hunger is never fully objective. In the 1950s low blood glucose levels were supposed to be the main cause of hunger (this was the 'glucostatic' theory of the late Jean Mayer, a Harvard nutritionist). When spending the afternoon with a toddler, sometimes it feels as if you can see their blood sugar levels dropping by the second. If you catch them in time with a banana, all will be fine. If not, they will flatline into a tantrum.

Sure enough, there is a link between blood glucose and hunger. As we know from people with diabetes, our bodies react severely when blood sugar is too low: shakiness, nausea, sweating. But this does not mean that blood glucose can be used as a simple measure of hunger. It transpires that absolute glucose levels in the blood are actually a very weak indication of hunger and fullness.[11] Glucose injections do not tend to reduce appetite, much. High glucose in the blood, strangely enough, does not translate to high glucose in the small intestine.[12] What does induce hunger is when blood sugar rapidly drops. In a lab situation, a decline in blood glucose over a short time-frame – a few minutes – will fairly reliably prompt human subjects to ask for a meal. Yet this is still not the whole story. In one study, fifteen overweight men were asked to stay in an isolated room free of clocks or any other indication of when mealtimes might be, while their blood sugar was continuously monitored.[13] They were free to ask for a meal whenever they liked. Between them, they asked for forty meals when they were in a 'postabsorptive state', meaning when all the food they had eaten at their last meal had been absorbed from the digestive tract. Yet in these men, these hungry moments never coincided with a drop in blood sugar.

Another 'biomarker' for hunger is a hormone in the gut called CCK, which is released into the blood when the body detects fat or protein. At least sixteen separate scholarly studies have shown that the presence of CCK suppresses hunger, but since it works primarily when the stomach is full, it hardly represents much progress as a measure for hunger. After twelve healthy men were given slow intravenous doses of CCK, they spontaneously reduced the amount of food they ate (including strawberry jelly on Ritz crackers) by an average of 122g. But that average conceals great variation. Three of the twelve men actually ate *more* food when they were dosed with CCK. A 2003 study found that the appetite-suppressing effects of CCK are greatly increased when the stomach is distended.[14] But we already knew that people with distended stomachs tend to feel full, CCK or not.

There has been much excitement in scientific circles over two other hormones, leptin and ghrelin, that appear to work in tandem to influence our hunger. The theory is that leptin reduces hunger and ghrelin increases it. Certainly in animals, high levels of leptin – a hormone that instructs the brain how many body fat stores are available – make them eat less. With humans, it's complicated. There have been rare cases of hugely obese children whose bodies lacked leptin and who were crazed with hunger, scavenging rotting food from rubbish bins or uncooked fish fingers from the freezer. Leptin injections got them back to normal levels of both hunger and weight. Among those of average weight, however, leptin concentrations do not change markedly after a meal; it is only after starving for longer periods, twenty-four hours or so, that leptin concentrations in the body fall significantly.[15]

Given that leptin sends information about how much fat is available to the brain, its usefulness as a measure of hunger depends on what else is happening in the body. Binge eaters often have relatively high levels of leptin circulating in their system, as do obese people, but this does not mean that they never experience hunger.[16] It has been suggested that obesity may trigger leptin resistance in the body, such that it no longer works as a signal for reducing appetite.[17]

Ghrelin – a hunger stimulant – may be a more promising biomarker for hunger. People who suffer from the rare condition Praader-Willi syndrome, characterized by extreme and unassuagable hunger, have 4.5 times more ghrelin in their system than others. But the absence or presence of ghrelin is not enough to cause hunger. In people with regular mealtimes, the feeling of hunger appears to arrive *before* the amounts of ghrelin in your system increase.[18]

The most common way scientists measure whether someone is hungry or not remains simply asking them. Your answers are then mapped on a scale. For example, you might be asked, 'How hungry are you?' and prompted to mark your answer on a line somewhere between 'not at all' and 'as hungry as I have ever felt'. Our body gives us some pretty clear signs that it wants food, including tightness in the stomach, loud rumbling of the digestive tract, a feeling of hollowness, light-headedness, dryness in the mouth or throat and sometimes a weird excitability. In pigeons who have had the cerebrum in their brain removed, hunger leads to an intense restless running about, which instantly stops with just a few grains of wheat.[19] Humans are not pigeons. The problem with subjective reports of hunger is that different people experience lack of food in vastly different ways.

My mother was a wartime baby – born in 1941 – and many times told me and my sister of the hunger of the late forties, after the war, when rationing was still in force. She attributed her lifelong fear of small portions to this austerity. Rationing had taught her that no portion could ever be too big. Years later, on long family car journeys, she could not bring herself to stop at the motorway restaurants called Little Chef, not because she thought the food was bad – which it was – but because the adjective 'little' convinced her there wouldn't be enough of it. A story she told repeatedly to illustrate the hunger of rationing was of the day as a child she felt so ravenous that she burst into the larder and ate a whole block of margarine. Such hunger! But one day, her older brother, my uncle, was having lunch at our house when she told the margarine story again. Did he remember feeling terribly hungry too during those wartime years? she asked. Not particularly, no, he said, and changed the subject.

The most famous attempt to measure what hunger actually does to the human body was the Minnesota Starvation Experiment, conducted in 1944–5 at the University of Minnesota.[20] For twenty-four weeks, thirty-six strapping and healthy young men were put on a reduced diet of 1,560 calories, a similar amount to many weight-loss diets now. As well as losing on average a quarter of their body weight, they suffered intense psychological and physical distress. Some became obsessed with reading cookbooks. Many found that their sex drive declined and they withdrew socially. It was common to feel dizzy, moody and nervous. They bit their nails, chewed gum and drank coffee in profusion. A few became paranoid that they were being given less food than others or engaged in bizarre behaviour such as dousing their meals in water and spices.[21] But while they all ate the same amount of food, they varied in the degree and quality of the hunger pains they felt. Two-thirds reported feeling hungry all the time, but another third did not. For some, hunger felt like a mild discomfort in the abdomen; for others it was a sharp and intolerable shooting sensation. In the months after the experiment ended, their ability to gauge hunger went haywire. They might eat as many as 6,000 calories, gorging until they were uncomfortably full and gassy, yet still feel unsatisfied.

Hunger – this mechanism that we suppose to be so basic – turns out to be one of the more intricate bodily impulses. To feed our hunger is not like putting petrol in a car. It is not straightforward to gauge its extent, either from the inside or the outside. Nor is it an easy thing to terminate. The mere fact that we speak of hunger as simple may be a sign of how little we have understood what it would mean to master it.

'You'd eat it if you were really hungry,' I hear myself saying to the child who is pushing the remains of the mashed potatoes around on his plate, angling to be offered something better. 'If you were really hungry, you'd be happy with a slice of bread,' I snap at the ravenous teenager who has already eaten a full supper, plus pudding, plus a supplementary bowl of yoghurt, fruit and honey, plus a toasted cheese sandwich and now says he can't sleep until he has one last snack. I hate the naggy way my voice sounds when I say this 'if you were really hungry' line – which I only come out with at the end of my tether. It's as if I am blaming my children for not being more like those other more deserving children in the world, the 'really' hungry ones. The implication is that 'if you were really hungry', you'd eat anything. Which isn't actually true.

What is true is that hunger teaches people to accept a wider range of foods. In one experiment from 2009, people were either deprived of food for fifteen hours or given plenty to eat.[22] After, when they were shown pictures of unappetizing food – stews that were reminiscent of vomit, spinach pulp – the hungry people showed less activity in the levator muscle of the mouth that signals disgust. Conversely, when shown pictures of appealing food – pasta, pizza – the hungry subjects showed more activity in the zygomaticus muscle, which makes us smile. Hunger made the nasty food seem nicer and the nice food seem nicer still.

But there are limits. Even in a state of abject food deprivation, there are certain taboo things that you would not eat – witness the fact that cannibalism is so rare. Hungry or not, few in the developed world would resort to a dinner of insects; or eyeballs; or dog. In most situations involving unusual foods, disgust easily trumps hunger. It is false to think that there is a state of absolute hunger in which children would eat anything. Among the hungriest children of the world, hunger is still concrete and not abstract. It cannot be satisfied by just any old thing.

Over the past ten years, the treatment of acute child malnutrition – the sort of hunger that carries an imminent danger of death – has been revolutionized by the invention and distribution of a peanut-based paste called Plumpy'Nut.[23] This is an energy-dense mixture: a kind of super-charged peanut butter, delivered in little foil-wrapped packages that children can squirt straight into their own mouths. It was the brainchild of André Briend, a French paediatric nutritionist, who came up with the idea after trialling numerous less successful malnutrition foods including doughnuts and pancakes. Briend supposedly had his 'aha moment' for a nourishing paste by looking at a jar of Nutella chocolate spread. On a trip to Malawi, where both peanuts and hungry children are plentiful, he borrowed an electric blender from a local restaurant and whizzed up a nutty cocktail of peanuts, milk powder, vitamins, minerals, sugar and oil.

Before Plumpy'Nut, when a child under five arrived at a feeding centre with suspected acute malnutrition, the safest option was to admit them to hospital for tube feeding. A high percentage – as many as 75 per cent in some centres – died anyway. Because it was so hard for mothers to be separated from their children, they often delayed bringing them in for help until it was almost too late. Another option was to give families a dried fortified milk mixture called F100 that could be administered at home, but that had to be diluted with water, a dangerous proposition in most of the developing world due to the lack of reliably safe drinking water. Also, the dosage and dilution of the milk powder was left up to the families and many over-diluted it, to make it go further, so that it could be shared among all the children in the family rather than just those at high risk of death.

One of the great advantages of Plumpy'Nut is that as a paste it does not have to be diluted, so it can safely be given at home, avoiding the need for a hospital stay. It is what is known as an RUTF: Ready to Use Therapeutic Food. The first trials with Plumpy'Nut yielded results that were miraculous. Mark Manary, an American paediatrician working in Malawi, field-tested it in 2001.[24] In defiance of medical orthodoxy, he sent all the children on his ward home with a six-week course of peanut paste. Ninety-five per cent of them made a full recovery, as against an average of just 25 per cent of those being treated for malnutrition in hospital. Six months on, the Plumpy'Nut children were still healthy. Peanut RUTFs are now the main way of treating acute child malnutrition around the world. In African countries these sweet pastes are popular with both children and mothers. Children like the sticky nutty taste. Mothers like the convenience. And doctors and aid workers like the fact that rates of recovery are so high.

And yet Plumpy'Nut has not been so welcome everywhere. While it has been an unequivocal hit in Africa, in India and Bangladesh the mothers and children do not respond to it so well. It's not that they are not hungry enough to appreciate it. There are around 8 million children in India at risk of death from severe acute malnutrition. Bangladesh, too, has one of the worst levels of childhood hunger in the world with 46 per cent of under-fives stunted and 15 per cent wasted.[25] As we saw in relation to girls in India, stunting means not growing or developing properly due to years of poor nutrition. Wasting is the other kind of hunger, the acute kind that can kill in a matter of weeks, brought on by sudden food shortages or disease: exactly the kind of critical hunger that Plumpy'Nut is designed to combat. But in Bangladesh a peanut paste does not fit with local ideas of what 'food' is and specifically which foods will appease a child's hunger.

Jose Luis Álvarez Morán works for the charity Action Against Hunger, which intervenes against child malnutrition in more than forty countries. Álvarez Morán has witnessed first hand how Plumpy'Nut has improved the treatment of acute malnutrition in most of the countries where the organization works. But not in India and Bangladesh. No matter how extreme the hunger, our cultural ideas about food do not go away. Indian mothers, on the whole, would rather feed a hungry child something made from lentils or rice than peanuts, which are not part of the everyday diet. 'And in Bangladesh, they just don't like it,' says Álvarez Morán. 'They want only locally produced food.'[26]

When researchers went into an urban Bangladeshi slum in Dhaka in 2011, they found a very low level of acceptance of Plumpy'Nut among parents and children.[27] If it were true that a 'really hungry' child would eat any food, then the slum-dwellers of Bangladesh should be only too happy to receive free sachets of calorie-dense Plumpy'Nut. But this was not the case. Out of 149 Bangladeshi caregivers of malnourished children – mostly mothers – six out of ten said that Plumpy'Nut was not acceptable as a food. Many hated the peanutty smell; others reviled the sweet taste and the sticky thick texture. The dark brown colour looked like excrement to three of the parents. Twenty parents said their child needed encouragement to eat it and fifty had to be forced. It was as if they refused to accept that this strange brown paste – so unlike food as they knew it – could be satisfying their child's hunger. Thirty-seven per cent said it made their child vomit and 13 per cent said it gave them diarrhoea, even though 112 of the parents also admitted that their child was gaining weight while consuming it.

This rejection of Plumpy'Nut is potentially a huge problem for charities such as Action Against Hunger who speak of a small 'window of opportunity' for reaching hungry children. Their task is not to dole out workhouse rations but to give a child a bank of nutrition at just the point their brain and body are developing most rapidly. To alleviate the hunger of a child, aid workers first have to reach the mother.

It starts with nourishing women before they become pregnant – another reason why the high prevalence of anaemia among girls is so detrimental. The window closes when a child is around two. If you can cancel a child's hunger during those first three years – from conception to toddlerdom – you create possibilities that stretch decades into the future. If not, the consequences can last for generations.

During the Dutch Hunger Winter of 1944–5, when the German occupying forces blocked food supplies, 22,000 people starved to death and a further 4½ million people suffered terrible malnourishment. Children born to mothers who were pregnant during that hungry winter – on rations of as little as 400–800 calories a day – were low birth weight and went on to have a host of health problems including diabetes and obesity. And their children in turn were low birth weight too. No matter what they ate later on, these people were blighted for their whole lives by the hunger their mothers experienced before they were even born. The power of early intervention with RUTFs – for both pregnant women and children – is to prevent this kind of 'irreversible harm'.

But it only works if the mothers and children will accept that the food will help their hunger. By the time someone is offering you Plumpy'Nut, by definition your hunger is far advanced and you need a lot of calories, fast. 'People forget,' says Jose Luis Álvarez Morán, 'that malnutrition is about disease rather than just lack of food. Many of these children have diarrhoea and actually need far more food than a child who isn't malnourished.' Contrary to the popular view, malnutrition is very seldom about an absolute lack of food. 'These families usually have farinas and porridges,' says Álvarez Morán, or rice and pulses in India and Bangladesh, 'but both the quantity and quality are lacking.' These farinas are rich in carbohydrates but deficient in essential micronutrients and protein. In this respect, they have something in common with the foods eaten by obese children in First World countries.

Development experts and pharmaceutical companies are now urgently looking at alternatives to peanut paste to assuage the hunger of children in India and Bangladesh. Some have suggested it would be more effective and sustainable to work with Bangladeshi mothers in the home to teach them how to make energy-dense foods – such as sweet sticky halva or milky puddings – rather than relying on commercial RUTFs made by big pharmaceutical firms. But the obstacles to mothers being able to cook the right food to treat acute malnutrition in the home – lack of clean water, kitchens or basic sanitation – were among the reasons that Plumpy'Nut was invented in the first place. There have been experiments with RUTFs based on local ingredients such as sesame seeds or chickpeas. The International Centre for Diarrhoeal Disease Research in Bangladesh is trialling some RUTFs that have 'a lot of potential', says Álvarez. They are made from rice, lentils, chickpeas, oil, milk powder and sugar and designed to have the same nutritional value as Plumpy'Nut.

The real test will be trials in the community. Will the mothers and children accept this rice, lentil and sugar mixture as something that might alleviate their hunger rather than making it worse? It is hard to say. A child's hunger cannot be cancelled by food per se. It matters very much what the food is.

One of the reasons hunger is so hard to pin down is that it is a negative concept, an absence. It is not-food, not-contentment. What we all want in our different ways is fullness: that blissful state where we do not desire another bite. But physical fullness has at least two aspects. The first is the short-term fullness that makes us decide a meal is over: satiation. The second is the longer-term fullness that tides us over during the hours between meals: satiety. When it comes to avoiding overeating, it is satiety that is the more useful. It is satiety that will – in theory, at least – stop you from impulse snacking or getting desperate for lunch half an hour after breakfast. Many dieters now obsess about satiety, searching out the meals that will deliver the maximum hours of fullness for the minimum calories. But as a child, I don't remember once choosing what I ate on the grounds that it would make me full in three hours' time. I was only interested in whether it would make me full now, this instant.

When we are young, our idea of fullness does not tend to look far ahead. The child's idea of fullness corresponds to 'satiation': that feeling of being satisfied that the meal is done. Children point to different areas of the body to show how full they are: up to their belly button, up to their neck, up to the sky over their head. In theory, satiation is what causes us to stop eating, though in practice it's always more complicated. When a family is short of food, we may stop eating long before we are satiated. When food is abundant, conversely, it can be hard to convince ourselves to stop, even after satiation is reached. We can be full and not full at the same time. Our waistbands may feel tight but that bowl of roast potatoes on the table calls to us. *Hara hachi bu* is a Confucian principle, popular in Japan since medieval times, that you should eat until you are only eight-tenths full. This principle has since been given backing by nutrition scientists who note that when we eat there is a time delay between the body receiving the food and the brain registering that we are full. When the urge comes to have a second helping, it's worth waiting twenty minutes, and the feeling may pass. If we carry on eating until we are full up to the sky, then we will, in fact, be overfull.

All over the world, children gesture that a meal has satisfied them by rubbing their rounded tummies. This means: I am full. Children are correct to think that satiation mainly happens in the stomach. When food makes its way to your stomach, the vagus nerve tells your brain that you are starting to feel full.[28] The feeling of 'distension' in the stomach is a crucial element in satiation. This is one reason that weight gain is so hard to reverse – particularly when caused by binge eating – because the obese have increased gastric capacity, meaning that it takes longer for the stomach to feel full. And if our stomach doesn't feel full, the brain can't feel full either. One study found that when subjects received an infusion of tomato soup through a tube to the stomach, it gave them a feeling of fullness whereas when the soup was administered to the intestine, it did not satisfy to the same extent, though the body was receiving the same nutrients. As nutrients are released from the stomach to the intestine, however, the brain gets further messages from hormones in the gut that it is time to stop eating.

One of the commonest ways our eating goes wrong is that we consistently choose foods that offer immediate satiation in the belly rather than longer lasting satiety. When my eating was out of control, I remember thinking that a meal such as salad couldn't possibly be substantial enough. A childish longing for satiation leads us to stodgy foods – soft, pillowy cakes and buns, jam tarts, fluffy baguettes, buttery noodles, doughy pizza – which we imagine will stuff us like the plush filling in a teddy bear. When we crave them, much of the appeal is how replete we predict they will make us and indeed, in the moments after consumption, such foods do increase blood sugar fast. This appeal continues in adulthood, particularly for those on a low income. When you have little money to spend on food, it's hard to risk it on a bag of broccoli, which does not look as if it will dent your hunger, as opposed to a package of instant noodles or processed cereal, whose starchiness holds out the promise of quick satisfaction. It has been well documented that poorer households often gravitate towards food that is calorie-dense, low in fibre and high in carbohydrate and fat, stuff that looks like it will be 'filling'.

What these 'filling' foods are not so good at – paradoxically – is keeping us full for a long time. Highly refined starches and sugars give us a spike in blood glucose followed by a crash. When nutrition scientists talk about fullness, they tend to be less interested in immediate satiation and more interested in satiety: that slow-burn feeling of fullness that carries on after a meal and delays your next intake of food. The most effective foods to choose for longer satiety are exactly the ones that many children think they will not be satisfied by: high-protein foods, especially fish; soup; and high-fibre foods such as wholegrains and vegetables that are now sometimes said to have a low 'glycaemic index'. Low GI foods are ones that cause only a small rise in blood sugar. These include pulses and beans, salmon and eggs. High GI foods include white rice, sugary cereals, sliced bread – every kind of refined carbohydrate.

In 1994 a mother of an eleven-year-old despairingly told researchers conducting a survey on children's food, 'I just wish they'd eat more fruit and veg but all she seems to like is chips and biscuits – all the filling-in foods.'[29] The 'filling-in' foods have a universal appeal. The actress and cook-book author Gwyneth Paltrow, whose family follows a low-carbohydrate, high-protein diet majoring on leafy vegetables, pulses and fish, writes that every day her daughter begs for mashed potatoes, though they are forbidden.[30]

The search for a food that will keep us fuller for longer has been the Holy Grail of recent nutrition research. Most of the research has been done in labs using 'preloads' of various nutrients.[31] 'Preload' is a technical term for 'starter': something you eat before a meal that takes the edge off your appetite. Subjects are told to eat a preload of a particular food and are then monitored to see how it affects their subsequent energy intake and hunger levels. In these studies, protein emerges as one of the likeliest candidates for helping with satiety, more effective than either carbohydrate or fat. When people in a lab are given a substantial protein preload – whether it's tofu, meat, eggs or the whey protein used by body-builders – they generally seem to eat less at lunch an hour later. Another factor that may make a preload more filling is how viscous it is. Certain new fibres have been created that form a viscous gel in the stomach to hold off hunger for several hours.[32] The idea – which sounds a little creepy to me – is to drink an alginate liquid extracted from seaweed. On contact with stomach acid, the liquid forms a gel inside you, creating a sense of fullness.

Other studies have pointed to high-fibre grains as being good for satiety. A benefit of fibrous foods such as oats or crunchy fruit is that they take longer to chew, which gives the body time to register that it is properly full. That old warning shouted at children not to 'bolt your food' was sound advice. It's far harder to bolt a brown rice and kale salad than it is a white-bread ham sandwich.

Surprisingly, however, two of the most powerful ingredients to use as a preload for satiety are air and water, suggesting that fullness is not simply a matter of nutrients.[33] When a group of '28 lean men' were given preloads of milkshakes, they ate significantly less lunch when the milkshakes were whisked up with air. They felt fuller after a milkshake whose volume was 600ml as opposed to one whose volume was 300ml, even though the energy content of the two milkshakes was identical. It's possible that the increased air made their stomachs feel more distended. But the researchers concluded that 'seeing the bigger volume' of milkshake made them feel that they were consuming more.[34]

If you want to trick your body into feeling fuller for longer, one of the best ways is to eat soup.[35] When we consume liquid calories in the form of a drink – a sugary fizzy drink, say – they do little to fill us up, the supposition being that they pass through our mouth too quickly to signal to our pancreas, gut and brain that we are receiving nutrients. But if we consume liquid calories more slowly – spoonful by spoonful – and call it food, it becomes very filling, often more so than solid food.[36] Chicken soup, for example, was found in one study to be more filling than a grilled chicken breast. In another study, a starter of soup made people eat less of a main course than a starter of cheese and crackers, despite the fact that the cheese and crackers were more calorific. The filling properties of soup may be – as with airy milkshakes – partly to do with its large volume. There is evidence that – contrary to what our childish tummies might expect – we actually get more satiety from foods that are less energy-dense and soup fits the bill here, unless it's lobster thermidor or vichyssoise enriched with half a pint of double cream. We have a general tendency to eat roughly the same weight of food each day, regardless of its energy value, so it makes sense that watery soup should fill us up for fewer calories than other foods.

The filling properties of soup are not entirely rational, however. It is the *idea* of soup that makes us full, as much as anything. There's a reason why those books are called *Chicken Soup for the Soul*. They could just as well be called 'Pho for the Soul' or 'Clam Chowder for the Soul', because all soup is soul food. In pretty much every country in the world, something hot and brothy cooked in a pot and served in a bowl is viewed as uniquely nourishing. Soup places low demands on the eater. It treats you as a child, who may or may not know how to use a knife and fork. You do not have to chop, or even to chew. Soup is what our mothers gave us when we were ailing. It's what we return to after a hard day at work, when all we want to do is curl up in a foetal position on the sofa.

A good home-made minestrone prepared with bone stock, pulses, assorted vegetables, olive oil and pasta is a deeply nourishing meal, packed with fibre and protein, so no wonder it leaves you full. But our souls can be satisfied by lesser soups too. Like many working women in the 1980s, my mother resorted to packet soups. When she had the time, she could make a mushroom soup, heady with cream, parsley and shallots, as cosseting as a hug. When she was too busy, I was given a bowl of packet chicken noodle, consisting of potato starch, salt, flavourings and MSG with soggy pasta and tiny unconvincing shreds of 'chicken'. I looked up the nutritional values of packet chicken noodle and if it hasn't changed, my bowl of soup was a 43-calorie lunch: less than an apple. Yet I don't remember leaving the table hungry. It was soup – and just as importantly, soup being given in a loving spirit in a Bunnykins bowl at the kitchen table – so it filled me up.

Soup leaves us full because we believe it will. One of the

most interesting experiments on soup and fullness was done by scientists at Purdue University in 2004. They found that when apple juice was heated up and presented as 'apple soup' in a bowl with a spoon, it left subjects much fuller than when they drank it cold, in a glass as juice. The calories and the volume were identical, yet the apple soup left people much fuller than the juice both fifteen minutes later and an hour later. The researchers noted that soup may satisfy us more than juice because we perceive it as filling. They concluded that the main reason soup had such a positive effect on satiety was 'cognitive'.[37] We think of it as a food that will kill hunger; and so it does. The apple soup experiment suggests that to search for 'fullness' in this or that food is to look in the wrong place.

The search for special fullness-inducing foods assumes that when our bodies register that we are full, we will stop eating. This is a logical enough assumption, though, as we've seen, very little about the way we eat is, in fact, logical. Professor Barry Popkin, a leading obesity researcher based at the Carolina Population Center, argues that outside food labs, hunger and satiety are not actually the main forces driving our eating any more. Very few of us allow ourselves – or our children – to experience the sensation of hunger at all any more. We are semi-sated much of the time, preloaded with nibbles here and there that hardly seem to register as eating.

Popkin used survey data to track the way that meal patterns had changed in the US from 1977 to 2006.[38] He found that for both children and adults, there was a 23 per cent decrease in the amount of time between 'eating opportunities' (aka meals and snacks). In 1977 the average time between meals was 4.1 hours for children and 4.4 hours for adults. (For point of comparison, in eighteenth-century Europe a six- or seven-hour gap between meals was standard.[39]) By 2006, the time had gone down to 3.1 hours for children and 3.5 hours for adults. In other words, the time between 'eating opportunities' has shortened by a full hour. This is reflected in the rising number of calories eaten: in 1977 it averaged out at 2,090 a day, as against 2,533 in 2003–6 (across all American age groups over the age of two). Interestingly, the amount of calories consumed at meals has actually declined slightly for children since 1977 – by 62 calories a day. For both children and adults, the main increase in calories was from snacks. If Popkin's data accurately reflect eating patterns (and he suspects they are on the conservative side), the average adult now consumes an extra 180 calories a day from food snacks compared to thirty years ago, not to mention the extra calories from beverages (which increased from 290 calories to 422 calories for adults). 'Eating between meals' used to be frowned on, but now is actively encouraged by some diet gurus, who claim that our blood sugar will stay more regular if we snack every three to four hours. Which may make sense if the snack in question is a handful of pecans or a pear, but the argument starts to look a little shakier when you look at the snack foods that people most often buy: crisps, sugary muffins, confectionery.

Sometimes, the hunger won't go away, no matter how much you eat. It's easy to confuse hunger with other emotional states. When it's your birthday, you can't not be hungry for cake. 'Emotional eating' usually refers to sad emotions, but happy moods can also make us eat more. We are conditioned to use excess calories to celebrate. Researchers have found that it is possible to trigger celebratory binge-eating behaviour by putting subjects in a good mood, simply from watching a two-and-a-half-minute heart-warming film about a baby panda sneezing. In one test, the group who watched this film consumed 100 calories more of snack foods (M&Ms, peanuts and wine gums) than a control group who were shown a dull film about birds in the desert.[40]

Mostly, we eat so often and so much because we have more or less lost touch with the signals our body is sending us about hunger. We take our cues to eat from many places, very few of which have to do with the biomarkers of fullness in our brains and guts. Knowing which foods to eat, as we have seen, is a skill that develops with age and experience. But knowing how much to eat is something that infants are better at than older children or adults. Up until the age of three, children have a remarkable ability to stop eating when they are full.[41] It doesn't matter if you serve them a big portion or a small portion; they will eat until they are not hungry and then stop (assuming they are not force-fed). After that age, however, this ability to self-regulate hunger is partially lost; and sometimes never regained.

Young children's accuracy in recognizing when they are full has been confirmed by numerous studies. In one, from 2000, thirty-two preschool children from Pennsylvania were served macaroni cheese in two different age groups: three-year-olds and five-year-olds.[42] The three-year-olds ate roughly the same amount of the macaroni whether they were served a small, a medium or a large portion. They were not paying attention to the size of the food they were being offered but to what their own bodies were telling them. The five-year-olds, on the other hand, ate significantly more when the portion of macaroni was large. It was as if the sight of so much food in the bowl was telling them to ignore their own fullness and keep munching. The loss of hunger regulation after the age of four is a phenomenon that transcends cultures and continents. In 2013 the Pennsylvania findings were replicated in Kunming in China with variable portions of rice, vegetables and protein.[43] This time, the two groups of children were aged four and six. The four-year-olds from Kunming actually ate slightly less when offered a large portion, as if they felt overwhelmed by all that food. But the six-year-olds ate substantially more. The scientists behind this Chinese experiment suggested that 'in hunger and satiety, there is some point in the development process when children begin to respond to contextual cues such as portion size'.[44]

Most of us continue to respond to these contextual cues, rather than actual hunger, for the rest of our lives. In one famous study – the 'bottomless soup' experiment – adult diners were served tomato soup from bowls that were continuously refilled from secret tubes as they ate.[45] Others at the table ate from normal soup bowls. After the meal, researchers asked them how full they felt, and how much soup they believed they had consumed. Those eating from the self-refilling bowls estimated they had consumed just a fraction more than from a normal bowl: an additional 4.8 calories. In fact, they had consumed 76 per cent more soup than from normal bowls, nearly a whole extra portion. Yet they did not rate themselves as any fuller than those eating from the regular bowls.

From childhood onwards, our idea of fullness is heavily influenced by how much food we are offered. Large packages make it seem normal to eat large quantities. We are disposed to think that we will be full when we have eaten 'one' of something: one sandwich, one apple, one biscuit. Then, if we are extra-hungry, we might have a second helping. This was all well and good in the days when biscuits tended to be the diameter of a coffee cup, rather than a side plate. But the rise of vast portions – particularly in fast-food restaurants – means that if we eat only the calories we need, we should often stop at half of something; or even a quarter. And no one – child or adult – seems to like the feeling of the glass – or the plate – half empty.

My youngest son often demands not one but two biscuits, one for each hand. This is fine at home where I bake almond-butter crescents, not much bigger than a coin. But when out in a café, where the baked goods are giant, I've been known to break one in half and say, 'Look, now you have two.' Which doesn't fool him. Besides which, two halves of a giant cookie is still too much.

Professor Marion Nestle of New York University has spent decades decrying the 'Law of Portion size: the more food in front of you, the more you will eat'. One of her colleagues came into the office one day with the largest slice of pizza either of them had ever seen, measuring 14 inches long and weighing a full pound, equivalent to 2,000 calories: the full recommended daily intake for a moderately active woman.[46] The customers buying this pizza might tell themselves 'it's only one slice' and feel it must be OK to finish it: it's not as if you are eating a whole pizza. We need new eating methods to take account of the new ways we are being supplied with food.

The work of Brian Wansink, an expert in both marketing and nutrition, has shown that both children and adults can be disturbingly impressionable when it comes to deciding how much to eat.[47] We may believe we only eat until we are full, but there are countless triggers messing with the off-button for eating. Wansink has done a series of studies that involve manipulating the size of utensils to demonstrate what he calls the size-contrast illusion. A large bowl makes you eat too much ice cream; an oversize plate makes you serve too many potatoes; and a short, squat glass makes you pour too much juice. When judging the quantity of liquids, almost everyone focuses on the height and forgets about the width. This mistake is even made by experienced bartenders, who consistently overpour shots when using short tumblers as opposed to tall highball glasses.

It's a chancy business, taking our cues about how much to consume from our surroundings. Wansink has found that sometimes just the sight of food is a powerful enough trigger to override sensations of fullness. When you have lost touch with the hunger signals of your own body, the prompts to eat are almost inescapable. We are like Alice in Wonderland, controlled by cakes that say Eat Me and bottles that say Drink Me. When dieters were asked why they stopped or started eating, some of them simply said, 'I saw the food.'[48]

Countless studies have shown – duh! – that we eat more when distracted by a screen, whether TV, tablet or computer. A study of nine- to fourteen-year-old boys showed that not only did they eat more while watching TV, but the larger quantities of food did not make them feel any fuller.[49] What was happening on the screen was way more interesting to these boys than what was happening in real life.

Wansink has laid out some simple ways by which we can re-engineer our food environment to eat less. Avoid 'distracting' meals such as TV dinners and computer lunches. Replace the biscuit jar with a fruit bowl. Repackage food into smaller containers. Order half-size portions in restaurants. Replace short wide glasses with tall narrow ones. And get smaller plates. This last one has certainly worked for me. Sometimes I know I'm not hungry at the end of a meal, but yearn for something very sweet to punctuate my eating. I get my tiniest plate – the blue and white china kind you buy from Chinese supermarkets for dipping sauce – and fill it with whatever I crave: dark dense chocolate cake, vanilla ice cream with caramelized almonds, sticky gingerbread. It doesn't matter how full I make the plate because it will still be a tiny portion, so I can eat it without guilt or remorse. The first time I did this, I was sceptical: could I really be so childish that my brain would be fooled by the smaller plate? Yes. I could.

Another cue that keeps us eating more than we should is variety. When asked why they stopped eating, participants in food studies are just as likely to cite boredom as fullness. The waiter's insistence that we have an extra stomach for sweet things turns out to be true. Sort of. Professor Barbara Rolls, working at Johns Hopkins University, coined the term 'sensory-specific satiety' meaning that as we eat a certain food, our hunger for that particular food declines; but our hunger for other new foods remains fresh. This is why buffets are so dangerous. Just as your hunger for one food wanes, there is always something else there to tempt you to eat more. Rolls has argued that the original evolutionary purpose of sensory-specific satiety among our hunter-gatherer ancestors was to promote a good varied diet.[50] But it does not work so well in our modern food system where variety might mean different colours of sweets or different flavours of popcorn.

If mindless eating is what makes us blind to our own fullness, the solution might be found in mindfulness, that current new-age buzzword. Training in mindful eating teaches you to pay greater attention both to the food and to the sensations in your own body. Before you sit down to eat, you ask whether you are really hungry. You set the table nicely, with candles and napkins. You switch off distracting electronic devices. You take the time to savour the aroma and the flavour of the food, putting down your fork between mouthfuls. You notice whether you are enjoying the food or not and if you are not, you stop eating. Obviously all this is a little tricky to achieve if you are a child sitting at a table with a parent bellowing in your ear to finish your breakfast or else you'll be late for school.

The work of paediatrician Susan L. Johnson, however, has shown that it is possible to teach children to become better at responding to their own internal fullness.[51] Many parents, especially those who themselves struggle with their weight, believe that a child is incapable of self-control when it comes to eating. They themselves may be so out of touch with their own hunger cues that they do not credit their child with the capacity for learning to eat only when they are hungry. Yet twin studies suggest that the ability to stop eating when you are full has minimal genetic cause: it is fundamentally a response to environment. It can therefore be learned.

Johnson demonstrated that over six weeks of intensive intervention, it was possible to train children to improve their ability to self-regulate the amount of food they ate. The children were in a nursery and had an average age of four to five: just the stage when our natural ability to self-regulate portions deserts us. When Johnson first assessed the children, they varied wildly in their hunger regulation. Some overate, some underate and some 'regulated accurately'.[52] The children whose mothers dieted and had difficulty managing their own food intake were the ones who were least skilled at regulating what they ate in response to hunger.

Johnson and colleagues used doll play to help children start to recognize whether they were hungry or not, as well as talking about hunger and overeating and talking about what it feels like, from tummy rumbling to an uncomfortably full belly. Special dolls were brought into the nursery with nylon 'stomachs' filled with varying amounts of salt. Some of the dolls had empty stomachs that were 'hungry'; some were a little full; and some were very full. At snack time, the children were asked to place their hands over their own stomachs to see how full they were and to choose the doll whose stomach felt most like their own. By the end of the intervention, children had started spontaneously to say things such as, 'I'm not hungry any more so I'm going to stop eating.'[53] At the end of the six weeks, there were 'significant improvements' in the way that the children ate at snack time. The overeaters started to eat less and the undereaters to eat more. Johnson's study suggests that with the right teaching and support, 'children are capable of controlling the *how much* of eating'. The first stage is learning to recognize whether our stomach is empty or not.

This is something adults can learn too. In one weight-loss trial, it was found that giving participants lessons in mindfulness about hunger resulted in greater reductions to BMI than dieting alone.[54] Another seven-week intervention with obese and overweight adults in the Netherlands trained them to reduce their cravings by accepting them.[55] When these people felt assailed by food cravings, they were taught not to eat immediately but to wait and observe what they were feeling, both in their bodies and minds. The psychologists doing the intervention gave the patients techniques to become more aware of their own hunger, and to observe whether it was physical or emotional. They were taught how to perform a 'body scan' on themselves, methodically paying attention to different parts of the body and gauging for signs of hunger or satiety. This is not so different from playing with a doll who has a stomach full of salt. At the start of the seven weeks, many of the participants believed they had no command over their own hunger. They would say things such as, 'I cannot stop eating until the bag is empty.' The training taught them ways to experience the various hungers their bodies felt without being controlled by them. They learned to accept their cravings without always giving in to them.

The latest January diets will often claim that if only you follow all the steps, you will never feel hungry again. It's taken me a long time to realize that part of eating well is making friends with hunger. We are not the starving children. To feel mildly hungry two or three times a day – when you are lucky enough to know that another meal is coming soon – is a good thing. All my life – except when I'd been attempting to lose weight – I'd responded to the gentlest of tummy rumbles as something that needed to be urgently cancelled out. It is only now that I see you can easily live with an hour or two of slight emptiness. In fact, it makes the next meal taste better ('hunger is the best sauce,' as the proverb goes). Eating without hunger and drinking without thirst can become so habitual that you forget how good it feels to regain the proper rhythms of feeding: to earn your meals before you eat them, even if all you've done to earn them is to wait.

It's like being a child, playing out on the street with your friends and losing track of time until your parents call you for dinner and you come inside, rosy-cheeked and famished.

BREAKFAST CEREAL

It is a ritual so ordinary we do not question it. Open box.
Pour cereal. Pour milk. Eat. Yet the breakfast cereal habit,
often acquired very early in childhood, can provide unhelpful
lessons in how to satisfy hunger. It teaches us that when we
feel empty, we should turn, not to a stove or to some whole
food we can choose on its merits, such as a piece of fruit, but
to a brightly coloured packet. Cereal is a medium through
which we learn to confuse hunger with marketing.

Ready-to-eat cereal has become the most ubiquitous
way to answer peckishness in the morning, last thing at night,
and plenty of other times in between. Boxed cereal – first
marketed as Granula in 1863 by Dr James Caleb Jackson –
was originally a quintessentially American custom. But the
rest of the world is catching up. In 2013 India consumed
$3.4 billion worth of breakfast cereal; Russia ate 3.8 billion;
Brazil 4.8 billion; and China 16.7 billion. Global sales of
breakfast cereal are increasing by around 10 per cent year
on year.

Breakfast cereal tends to be one of the first foods that
parents allow children to choose for themselves, as they sit
wedged in a supermarket trolley. Of all the foods that we
could give a child free rein over, cereal makes the least sense.

Parents let children choose cereal but dictate which vegetables they eat; it should be the other way around.

Cereal also tends to be the first meal we 'make' for ourselves, to fill any hungry patches in the day. Because it involves no fire or knives, cereal is seen as a 'safe' thing for children to prepare. The act of sitting at the kitchen table reading the back of a cereal packet forges an emotional bond that lasts a lifetime. The cheering lettering reassures us that, despite appearances, the sugar-coated flakes will do us good, because of all those added vitamins and minerals.

The cereals marketed for children are consistently the most sugary and highly processed in the whole cereal aisle. A child's choice isn't determined by the sensory properties of the cereal – which are hidden inside the box. You pick the packet that promises you the most stuff: the most bizarre shapes, the coolest cartoon character, the best free gift or promotion, the most added chocolate. This offers a bad lesson in how to choose food. When you encounter a genuinely filling breakfast food, such as oatmeal or scrambled eggs, it is hard to shake off the feeling that there is something missing.

CHAPTER 7:

Disorder

Rebeca only liked to eat the damp earth of the
courtyard and the cake of whitewash that she
picked off the walls with her nails. It was obvious
that her parents, or whoever had raised her, had
scolded her for that habit because she did it
secretively and with a feeling of guilt . . .

GABRIEL GARCÍA MÁRQUEZ,
One Hundred Years of Solitude

Diane was a forty-eight-year-old woman with a full-time job
as an office manager, with no outward signs of medical problems, aside from being overweight. She had never sought or
received professional treatment for any kind of eating disorder. Yet in 2014 her case came to the notice of researchers
studying adults in a deprived borough of the Midlands in
the UK who identified themselves as 'picky eaters'. Diane
was one of the participants recruited through local libraries
and leisure centres. In all, the researchers spoke to twenty-six
families, interviewing them at length in cafés and at home.

The 'picky eaters' were also asked to keep a detailed photo diary of everything they ate over a four-day period showing where they ate it, and with whom. This exercise built up a picture of apparently normal people who nevertheless ate in deeply abnormal ways. All the adult picky eaters said they had been eating this way since childhood. At the age of nearly fifty, Diane's diet consisted mostly of cheese, processed potato products, sliced bread and cereal.[1]

Despite having the skills and confidence to cope with her job, Diane was passive and childlike about food. She spoke of guilt about the way she ate and still felt she had let her mother down by not being more accommodating. Diane's eating was limited not just by ingredients but by temperature. The only vegetables she could tolerate were very cold salad vegetables, sliced thin with no dressing, in minuscule quantities.[2] Cooked food had to be very hot or she would not touch it. She went with one of the researchers to a café where she ordered egg and toast but had to stop eating it halfway through because when it got cold it 'turned her stomach', she said.

The way Diane ate made it difficult for her to go and eat at the houses of friends as she was apt to refuse the food they cooked for her and couldn't prevent the look of abject disgust on her face. Diane recognized her eating to be a problem; and yet from her point of view it would have been more problematic still for her to try to consume any of the foods that she did not want to eat. She hated cooking, she said. Occasionally, she would attempt to eat a more balanced diet but she couldn't bring herself to 'crave' anything but 'junk'. 'Well, I'm forty-eight now so I'm not going to be able to change am I, really?'[3]

This case does not accord with what we imagine to be

a typical eating disorder. Diane was not a teenager, she didn't – so far as we know – obsess over fashion magazines or do ballet, and she wasn't overly preoccupied with having a flatter stomach; indeed, she recognized that she would be healthier if she could bring herself to eat different foods. Diane was not avoiding food as a way of limiting her calorie intake but because she just couldn't bear to eat certain things. Yet there is no doubt that her life – and her health – suffered profoundly because of her eating. Diane's case shows just how distorted eating can become and still not quite count as an 'eating disorder' in clinical terms.

There are many misconceptions about eating disorders. One is that they are all about trying to lose weight, like anorexia or bulimia. Another is the view that they have nothing to do with the rest of us. In fact, eating disorders – whose forms are as numberless as snowflakes – are best understood as extreme versions of the dilemmas and pitfalls that all of us face in learning how to eat.

Those whose eating disorders are so acute that they – or their parents – seek help from an eating or feeding disorders clinic are the tip of an iceberg that extends deep into our food culture: a far-reaching edifice of low-level unhappy eating that manifests itself in countless ways. Think of the office worker who compensates for a hard day – every day – with comfort snacks in front of the TV; or the child who is scared to drink anything that doesn't taste sweet. There are the grown men who still feel sickened by greens, and grown women who won't allow themselves to order a dessert, only to eat the equivalent of a double portion from everyone else's plates. Most ubiquitous are the yo-yo dieters, stuck on an unremitting treadmill of losses and gains, who own two separate wardrobes, the fat clothes and the thin ones.

Such behaviour is so common that we do not recognize how dysfunctional it is. In a survey of 2,000 college students in the United States, 41 per cent of women and 18 per cent of men said they were currently 'on a diet'.[4]

Sometimes, in a black humour, dieters say that they envy anorexia sufferers their 'discipline'. The difficulties faced by someone who has pared down their consumption to almost nothing seem far removed from the more everyday disordered eating of the rest of us, and so we treat eating disorders as irrelevant. Who would choose to spend any time thinking about laxative abuse or starvation-induced organ failure when we could be dreaming about cake? We avert our gaze from anorexics in the gym changing rooms, not wanting to draw attention to their jutting collarbones and gaunt legs, though whether we are sparing their feelings or our own is not clear. Yet if we look closer – and preferably make eye contact – we might find that eating disorder sufferers have something to teach the rest of us about food. For one thing, they show just how high the stakes become when eating goes wrong. More importantly, eating disorders offer a model for how eating habits can be relearned from scratch, even by people whose genetic disposition makes eating naturally more difficult.

Under the right circumstances and with the right kind of help, some people manage gradually to replace destructive food habits with ones that sustain and delight. Anorexia has terrifying mortality rates, with as many as 20 per cent of all sufferers dying prematurely. Anorexics have suicide rates fifty-seven times higher than the general population.[5] And yet most sufferers do survive and some even recover. In one study, researchers followed a large cohort of women with anorexia and bulimia for seven and a half years.[6] At the end of the time-frame, 83 per cent of those with anorexia had achieved at least a partial recovery and 33 per cent had managed a full recovery (recovery was defined as the 'absence of symptoms' for at least eight consecutive weeks). The bulimia patients had even better rates of improvement, with 99 per cent enjoying at least a partial recovery and 74 per cent a full recovery. Think what this means: the vast majority of those whose meals had once been a maelstrom of bingeing and purging had now succeeded in learning how to eat food and digest it in a normal way.

The first step in recovering from an eating disorder is recognizing that there is a problem. There are more people than we might imagine who have an eating disorder without fitting the expected model of anorexic teenage girls. I have a friend whose childhood in the 1970s was blighted by an untreated eating disorder. At the age of around seven, after the birth of a sibling, family meals started to make him nauseous and he would spend several days each week off school, vomiting. His weight plummeted. But the local hospital pronounced that his symptoms were 'only' psychosomatic, not physical. His parents interpreted this to mean that he was making it up and did nothing further to help him. After his parents divorced, he slowly recovered, by himself.

The ways in which eating can go wrong in early life covers a rich spectrum, spilling out beyond the official categories and definitions. Anorexia is not in fact the most common childhood eating disorder. Nor is it bulimia. That honour goes to EDNOS: Eating Disorder Not Otherwise Specified. 'Other', for short. Our fertile minds are very good at concocting our own idiosyncratic eating disorders, with a bit of binge eating here, a smattering of purging there and skipped meals in between.

Some children are morbidly obsessed with certain foods; others have trouble summoning much of an appetite for anything. Eating may go wrong because a child is hypersensitive to touch – for some, lumps in the mouth are actively painful. Others are hypersensitive to smell: the odour in the school dinner hall may make a child feel unable to eat. There are children who can't swallow anything that isn't soft and those who vomit at the thought of a new food. Some (this is called pica) eat things that are not food – ranging from dirt to baby powder. Some chew food repeatedly and regurgitate it (this is rumination) and then decide whether to chew the regurgitated food again or spit it out. One ruminant explained his thought process: 'If it's a good slice of pizza, I'm not going to waste it. But if it was spinach to begin with, of course I'm going to spit it out.'[7]

Problems with eating can be placed in one of two big categories. There are the 'feeding disorders' of little kids and the 'eating disorders' of older children and adults.[8] Anorexia, for instance, is a classic eating disorder whereas a classic feeding disorder would be some kind of phobia of food or extreme pickiness. Too often, feeding disorders aren't taken as seriously as eating disorders. Until 2013, the guidelines in the official manual aimed at clinicians who treat mental health said that it was only a 'feeding disorder' if the onset was in a child younger than six. The idea was that pickiness – even extreme pickiness – is a developmental stage, something that the child will grow out of as they get older. A feeding disorder can look like a form of extreme childishness – and indeed some of the forms it takes involve children who won't consume anything but milk or baby food. An eating disorder, by contrast, involves intense preoccupation with one's body weight, like a warped version of the adult activity of dieting. It is often associated with depression, anxiety, OCD and suicidal thoughts.

It isn't true, though, to say that 'feeding disorders' are only for infants and 'eating disorders' are for teenagers and adults. To study disorders of eating is to see that we live in an era where some children are old before their time and some adults are for ever juvenile. There are prepubescent children – as young as six or seven – who have already started to limit their food intake to dangerously low levels because of fears that they are not 'skinny' enough. Equally, there are grown-ups holding down jobs and paying mortgages who will eat nothing but canned baked beans – the so-called 'adult picky eaters', most of whom slip under the radar, never treated by the medical profession.

Whether they affect adults or children, eating disorders and feeding disorders are two very different types of condition. Those with an eating disorder may be scared of eating a French patisserie such as a pain au raisin, rich buttery layers of pastry, sugar and fruit, because they think that eating it will make them fat. Someone trapped in a habit of selective eating – the main kind of feeding disorder – may also be scared of a raisin pastry, though they are not scared that it will make them fat. They are scared of the raisins themselves: the sight, smell, texture, colour and, above all, the very thought of them.

There are certain things, however, that the vast spectrum of eating and feeding disorders have in common. While the causes and experience of these conditions are very different, the treatment can be remarkably similar. Structured meals, at which a wide range of foods is offered in a firm but loving way, are a key part of the cure for any eating disorder (along with various talking cures and medication to help with anxiety or depression). Family meals, used to 'refeed' the child, are now the gold standard for treating childhood anorexia (assuming the patient is not in need of hospitalization). Likewise, the treatment for feeding disorders involves slowly building up a child's repertoire of foods through repeated tasting sessions. The whole family will probably need to rethink portion sizes (smaller for binge eaters and larger for anorexics) and where and how they eat. Often, an eating disorder has become so difficult for the rest of the family to manage that the child gets into the habit of eating alone. Recovery happens when the child is happy to be included in the social life of mealtimes; and when the rest of the family is happy to have them there. Whether the disorder is anorexia or restrictive eating, recovery involves learning how to eat all over again.

For people with eating disorders, food is both the poison and the remedy, but this is true for the rest of us as well. Food is an inescapable fact of life and the task for each of us is to find a way to make our peace with it. Disordered eating is very different from alcohol addiction, whose cure is sobriety. When eating goes wrong, the antidote is not a life without food, but figuring out how we can bring ourselves to eat new foods in new ways.

It is one of the great clichés about eating disorders that they aren't really 'about' food. This is a fairly senseless remark. You might just as well say that hay fever isn't really 'about' pollen. It's certainly true that there is far more going on in anorexia and bulimia than just the surface eating behaviour: starving or purging are a physical manifestation of some deep depression or cognitive malfunction. These are mental illnesses and the causes, as we will see, are as much genetic as environmental. But by the time someone is in the grip of anorexia, it is very much 'about' the food: the devouring of cookbooks, the meagre meals of a single piece of fruit, slowly eaten, the way a brain changes in the absence of nourishment.

Selective eating is first, second and last about the food. It's about the distress of being offered orange food when you only eat yellow. It's about the subtle differences between the brand of yoghurt you can tolerate and the one that makes you gag. Some children are so sensitized to the smell of the 'wrong' foods that they cannot even sit with family members when they are eating it. Because this kind of extreme eating usually goes along with other underlying conditions such as autism, coupled with behavioural difficulties at the dinner table – terrible tantrums and rage or sadness – it is easy to assume that a failure to eat all but a narrow range of foods is primarily a way of acting out some deeper problem, and that it would be trivial to pay too much attention to the eating itself. Yet the evidence points in the opposite direction. Whatever their core condition may be, these children cannot behave at the table because the food is causing them such distress. If people kept serving you dinners that made you retch, wouldn't you be angry too?

The great discovery of the past twenty years in the treatment of both eating disorders and feeding disorders has been that addressing the eating head-on offers the best chance of recovery. When therapists successfully manage to train selective eaters to enjoy a wider range of food, parents report that the children's behaviour at mealtimes also dramatically improves, even for children on the autistic spectrum. Food is never just 'about' food. The trials of being a selective eater – or of parenting one – bring home the ways in which food itself can represent so many other things in a person's life: control and worry and isolation and feeling safe. If you can improve the eating, the rest of life gets a little better too.

When you are a selective eater, your days are blighted by your need to skip situations where you might be expected to eat something that you find impossible. As you get older, you may avoid travelling; or meeting up with friends, since so many social occasions revolve around food. You make excuses why you can't eat: you are not hungry, you have a tummy bug, you've already eaten. It's like people who grow up illiterate, who will often go to elaborate lengths to conceal that they cannot read.

For selective eaters, food can become the basis of major life decisions. A mother contacted a feeding clinic with concerns about her eighteen-year-old daughter. This girl was due to start her first year at university. She had made her choice of college not because it offered the best academic course for her future career, or because she liked the location, but for the simple reason that the student cafeteria served pizza twice a day, at lunch and dinner. She had taken the precaution of travelling to the campus to taste the pizza, to check that it was the kind she liked: generic plain pizza, without the taint of oregano or spice.

It's hard for others to comprehend that there are those whose diets are so restrictive they cannot even bring themselves to eat bread; or French fries; or pancakes. 'It's not unlike a phobia,' says Keith Williams, who runs the Penn State Hershey Children's Hospital Feeding Program, treating around a thousand children with feeding disorders a year. Williams and his colleagues see some children with a fear of food so great that if someone enters a room carrying a plate that they know contains something they have never previously tasted, they will spontaneously vomit or gag before they have had a chance to look properly at what it is.

Sometimes children have a fear of food that is linked to a specific terror of what will happen when they put the food in their mouths. They may have an irrational dread of poisons. Or there may have been a traumatic event that made them become sensitive about swallowing, for fear of choking, gagging or vomiting. It is particularly common to have anxiety about swallowing lumpy foods. Chewing skills develop most rapidly between the ages of six and ten months, but only when the child is given something substantial to chew on. With babies who are kept on smooth baby foods for too long, the chewing action is delayed and the child may then become hypersensitive to lumps. This is called 'oral defensiveness'. Experiments show that children who are only introduced to lumpy solids over the age of ten months are far more likely to show feeding difficulties as toddlers. It is very common for a one- or two-year-old to be wary of lumps, but for some children the fear actually intensifies with age. It is sometimes referred to as *globus hystericus* or lump in the throat. Sufferers – who are often mistaken for anorexics because they tend to lose so much weight – perceive solid food as a foreign body and can endure terrible spasms to the oesophagus when eating. This turns into a vicious circle of anxiety, where they avoid lumpy food so much that they find it even harder to deal with when it is encountered.

An inability to swallow is relatively rare, but a general anxiety about eating anything but familiar comfort foods is very common. The majority of those who are referred to specialist feeding clinics are kids with 'special needs', especially those on the autistic spectrum or those with oral-motor problems that make it harder for them to chew and swallow. One study of 700 children diagnosed with feeding disorders under the age of ten found that 86 per cent had an underlying medical condition, 18 per cent had a behavioural problem and 61 per cent had some kind of oral dysfunction.[9] However, you only have to look around to see that this problem is far more common in the general population than medical admissions might suggest.

Researchers asked nearly 500 American adults who had been called for jury service about how they ate. Around 35.5 per cent, over a third, described themselves as picky eaters.[10] Before we panic unduly at this statistic, it's worth noting that picky eating means different things to different people. Sometimes people use 'picky' to mean being discerning, a quality that can serve you well. When my eating was out of control, I wasn't picky at all. I never met a pastry I didn't like. Now, I've learned to be fussy about Viennoiserie, mostly saving my appetite for those that are exceptional. Becoming a food snob has its uses.

Other forms of picky eating are not so beneficial. Among the adults called for jury service, the self-styled picky eaters reported much higher levels of social anxiety and distress about eating than non-picky eaters. There are clearly significant numbers of people whose diets – and lives – are restricted by very limited patterns of eating. Keith Williams observes that 'tons of kids are selective eaters but don't come to a feeding clinic, because they have otherwise normal growth and development'. Parents only tend to seek specialist help for feeding either if their child is already in the medical system for other problems, or if the situation is so acute it has led to them becoming severely underweight. In Williams's experience, the problem of selective eating in children is far from limited to his clinic patients. In some places, it is close to becoming 'the norm' because so many children are not given the opportunity to learn to like a wide enough range of foods. 'We see families who serve pizza five or six or seven or ten times a week,' Williams says.[11] He despairs of the attitude that it doesn't matter if a child eats nothing but sugar-coated cereals and the occasional vitamin pill. Williams encounters this attitude in middle-class families as well as in those on lower incomes. 'The parents say, "He'll be OK," but there's just no mechanism for change. Why would a kid like that suddenly eat something else?'

We assume that over time our tastes will gradually blossom of their own accord, like a flower, but with selective eating the pattern is for tastes to get ever more closed. Selective eating makes a mockery of the view that eating well is something instinctive, natural or easy for humans. By the time they are seen by a feeding disorder clinic, selective eaters may be down to just a handful of foods. By far the most common safe foods chosen are carbohydrates, followed by dairy foods, meat, peanut butter and some forms of fruit and vegetables. A typical case is of a ten-year-old girl who ate only peanut butter sandwiches, cheese and tomato pizza and apples. Psychiatrists working with her found that 'Tracy said she wanted to try new foods but they made her retch.'[12] How could someone's eating reach such a state? The problem is that when parents make an intervention to widen a child's repertoire of foods, they get punished with some strongly aversive behaviour such as screaming, gagging or vomiting on the plate.[13] It is agony for the parent to watch as a child retches because they made the mistake of serving them a sandwich made from the wrong peanut butter. Next time, they make sure to buy the 'right' kind and the child's habit gets still more entrenched. Well-meaning outsiders may advise that a child will try new foods eventually if you take their favourites away and let them go hungry, but the kind of eaters Williams sees may hold out for as long as four days without eating, at which point they are in danger of needing feeding through a tube, which can be even worse for all concerned.

Traditional treatments for selective eating have often set the bar extremely low, focusing on getting the child to eat something – anything – rather than expecting that they could ever become accustomed to a full selection of normal foods. Even clinicians are often daunted by the level of resistance that some of these eaters show. These conditions are very hard to treat. The course of treatment will vary according to what other problems the child has – behavioural issues, or extreme anxiety, or an underlying physical problem with eating – but in many clinics, it will take the form of some kind of psychological intervention, plus nutritional advice, plus medical monitoring. Such an approach, while better than nothing, is unlikely to yield dramatic changes because it fails to address the eating itself.

A thirteen-year-old boy was brought by his mother to a feeding disorder clinic. He ate very little except for crisps, dry breakfast cereal and bread sticks, plus a single probiotic drink that his mother made him drink every day. He was pale, tired and with a very low weight and height for his age, which he was teased about at school. His mother described him as a 'lazy eater' and had more or less given up on offering alternatives since she did not want to waste money on food he would not eat. The boy was diagnosed as anxious. After a course of cognitive behavioural therapy (CBT, a talking therapy to help people change the way they behave) and some nutritional advice, he was encouraged to work on introducing 'one or two' new foods. By the time he was discharged, noted the clinicians, his diet was still 'far from extensive'. He would now eat yoghurt, fruit smoothie, French fries and he would take a multi-vitamin. His diet was otherwise unchanged and he remained anxious about food.[14] He still ate no vegetables or anything resembling a proper, protein-based main course. Yet the doctors who had treated him did not seem to think that his treatment was a failure. 'In many cases,' they concluded, 'it will be neither realistic *nor necessarily desirable* [my italics] to achieve an eating pattern without any avoidance or restriction.'

No wonder parents and children are fatalistic about their chances of recovering from selective eating when even the doctors who treat it sometimes suggest that a cure is unattainable. Yet in some cases, clinicians have managed to help children to overcome their selective eating more or less completely over a relatively short period of time: weeks or months, not years. Methods of treatment vary, but a common thread is that the most successful interventions start with the assumption that selective eating *is* about the food, and that it is both possible and desirable for a selective eater to find a less unhappy way of ingesting it.

A nine-year-old boy, Diego, presented for treatment at a clinic in Sydney, after approximately seven years of selective eating.[15] He would only eat chicken nuggets, potato crisps and plain bread, to the exclusion of anything else. The food had to be eaten in a certain order and if the food appeared strange in any way – such as an unusually shaped potato crisp – the whole plate of food had to be thrown out and re-made from scratch. Diego often turned down birthday party invitations and failed to go to sporting events because he knew he just wouldn't be able to eat the food. Diego's eating was placing great strain on his parents' marriage, as his mother and father – who both ate a wide and varied diet – could not agree about how best to entice the boy into trying new foods. Each meal might last as long as three hours, during which time Diego became increasingly distressed and did not try one single bite of new food.

Diego's therapists realized they needed to find a new method of treatment that acknowledged just how scared he was by food. They helped Diego to give a name to his anxiety: 'Beaster the Worry Wart'. His parents could now tell Diego how sorry they were that Beaster was so hard for him to fight. Perhaps, they suggested, he should not try any new foods until they could 'figure out a way to tame Beaster'. This form of counter-suggestion is called a 'therapeutic paradox'. After seven years of cajoling and tears and stress at meal-times, this new tack must have come as a relief to Diego: now, he and his parents were on the same side in the fight against Beaster. The very next session, a week later, Diego arrived with a list of ten new foods that he had tried spontaneously, including steak and vegetables. Over a period of four months, Diego continued to try new foods until he reached a point of full recovery. Beaster, he said, had shrunk in size and was no longer so scary.

Such an approach might not work with every selective eater. An older child might find it babyish and a younger child might not be able to verbalize their terror about food in this way. A more obvious and universally applicable way to treat feeding disorders is intensive taste sessions to help change eating behaviour directly. If the problem is children not eating enough different foods, the solution is getting them to eat a wider range. Given the complexity of eating disorders, this sounds too simple, but this is what Keith

Williams and colleagues have done very successfully at Penn State Hershey Children's Hospital, where they have pioneered 'taste exposure' interventions to treat selective eating. Williams was familiar with Zajonc's concept of 'mere exposure' discussed earlier in the book. He knew that if you could get someone to taste enough new foods often enough, the odds are that they will start to like some of them. 'The trick,' he says, 'is getting them to taste the food.'

With standard outpatient treatment for selective eaters, parents may be advised to offer half a teaspoon of a new food at a family meal each day and ask the child to record their reactions in a food diary. There are at least two reasons why this might not work (setting aside the reluctance of small children to keep a diary). First, most parents have no training in how to offer novel foods. Quite understandably, they find it hard to ignore the tears and the rage and the throwing of spoons on the floor. When therapists observed parents who had been trying in vain for three months to treat their autistic children's feeding disorders at home, they found that the parents were only following around half of the recommended steps.[16] The children's attitudes at mealtimes dramatically improved after the parents were offered intensive training in how to offer new tastes, learning to ignore disruptions and quietly offer a new tiny bite when the first one is spat out. They were encouraged to be more ruthless about not allowing the child to snack on their own safe foods before a taste session, to give them a chance of being hungry for the tiny bite of new food. The size is very important.

When the contents of the spoon revolt you, even half a teaspoon may feel excessive. Keith Williams's clinic has seen positive results when the novel foods are first offered as small as a pea, or even as small as a grain of rice. If the food is tiny enough, and offered in structured taste sessions, even autistic children with extreme selective eating have been able to learn to like a wide range of foods, in under a week. In one intervention involving pea-sized bites given on ten consecutive days of treatment, three autistic boys were able to find an unexpected liking for fifty new foods. After just four days of treatment, they had more or less stopped their 'disruptive behaviour' at the dinner table.[17] Their parents were given training in how to continue the taste sessions at home.

Williams's latest version of taste exposure for selective eaters is called 'Plate A and Plate B'.[18] First, the parent chooses twenty new foods that they would like the child to try. Plate A contains three or four new foods chosen from this list of twenty, in pieces no bigger than a grain of rice (maybe carrots, chicken and oranges). Plate B contains foods that the child already eats without difficulty (let's say Pop-Tarts, biscuits and crackers). The parent gives the child four to six Plate A-Plate B meals a day, each lasting ten minutes (strictly timed with a timer) and no other meals are offered. They tell the child to have one bite from Plate A and then they may have a bite from Plate B and have a drink, 'ignoring crying or refusal'. The child continues to alternate between the plates until the time is up. When a child has learned to eat a food on Plate A for three consecutive meals without crying or gagging, the size is increased from a grain of rice to a pea; and then to half a spoonful; and finally a spoonful. By the time they are accepting a full spoonful, the child has learned to like the Plate A food. The aim is ultimately for as many as possible of the Plate A foods to become Plate B ones: something the child eats willingly and with pleasure.

The reason Plate A-Plate B can work so well – when

strictly applied – is that it places very low demands on the child. When food is as small as a grain of rice, it is almost as if it is not there. The pressure on the child is further reduced by the fact that there are several foods on each plate. If the child really can't stomach two of the foods on Plate A, they can always choose the third. Williams says that the reason this simple intervention is successful is that it gives children who cannot bear to taste new foods an opportunity to taste them. It helps them to vault over their own wall of resistance and put the food in their mouths.

Overcoming selective eating is even harder for older children and adults than for younger children, but it can still be done. Tyler was a sixteen-year-old boy with Asperger's syndrome.[19] His food restriction was so acute that for nine years he had been fed by gastrostomy tube. He had the height of a ten-year-old and the weight of a nine-year-old. Tyler ate just three foods: ham steak, cereal and pasta (which had to be bow-tie shaped: farfalle). Without the tube, he would not have received enough calories to survive. Previous attempts to improve his eating had been unsuccessful. Over a two-week course of treatment, therapists at Penn State Hershey created a modified version of Plate A-Plate B for Tyler involving a system of token rewards that he could put towards 'arcade' time on his laptop, DVD player and games consoles. For each meal, Tyler was asked to choose six foods, some easy and some 'difficult'. The more difficult Tyler considered a food to be, and the more bites he ate, the more time he could earn on his screens. The difficult foods all started off the size of a grain of rice and got gradually bigger. By the last three days of treatment, he was happily eating full-sized portions of normal meals: a main course, plus three or four side dishes.

By the end of treatment, Tyler had a repertoire of seventy-eight different foods, and several months after discharge was willingly adding new foods to his diet. He was now free of the wretched tube feeding. On cost grounds alone, Tyler's cure was a triumph: a year of tube feeding costs a minimum of $16,000 at 2007 prices, whereas his treatment cost less than $500 a day: $7,000 total. But the greatest gain was to Tyler's health and wellbeing. His parents reported that family meals were now actually enjoyable and he was gaining weight faster than he ever did on the tube. Tyler had left behind the loneliness of tube nutrition and was now enjoying the social interaction of a shared meal.

Keith Williams believes that, with the right motivation for change, it would be possible to use taste exposure to treat selective eating at any age. The greatest obstacle is that most selective eaters – and their parents – view their condition as incurable and therefore do not really believe that there is any point in treatment. Their reluctance in the face of new food is so great that they would rather organize their lives around the disorder – like the girl who chose the college that offered pizza twice a day – than fight the disorder and try to buy themselves a new life. It is even harder to treat selective eating in adults than in children. The adults may not cry and gag and spit, but they are less open-minded than children are about their potential to learn a new trick. Most would prefer to keep their condition as an embarrassing secret than contact a feeding disorders clinic.

There are exceptions. Over the years, Williams has worked with several adult picky eaters who desperately wanted not to be so limited and scared in their eating. When the motivation to change is there, he finds that taste exposure works just as well on adults as on children. He was once contacted by a

primary school teacher who wanted to become a mission-
ary in Asia but knew that she would not be able to make
it in a foreign country unless she could teach herself to eat
differently. At the time, she ate only ketchup sandwiches,
Oreo biscuits and instant noodles. Unless something changed,
she would never be able to cope with the food in the Far East:
it wasn't just the thought of the pungent flavours such as soy
sauce, ginger, spring onions and Sichuan pepper. She didn't
even eat plain white rice. Slowly and gradually, using taste
exposure and tiny morsels, she built up the foods she could
tolerate. That teacher is now working in the Philippines, land
of vinegar and garlic.

When your beautiful baby is first handed to you by the doctor or midwife and you see a whole future in their blurry
eyes, it's unlikely you'd imagine that they would become a
person who ate only ketchup sandwiches, Oreos and instant
noodles. What parents do spend a lot of time pondering,
particularly if the baby is a girl, is how awful it would be if
their child developed anorexia. What misery: to watch this
person you have fed so tenderly just waste away, rejecting
your meals and, by implication, your love. You would do
anything to avoid it, for your own sake as well as theirs.

Those of us who squandered too much of our youth on
stupid diets may feel especially determined not to 'give' our
child an eating disorder. We tie ourselves in knots trying to
save them from it. 'There's no such thing as perfect,' was my
mantra when my daughter was little and she scrunched up a
piece of her artwork. My fear was that her high artistic standards might seep dangerously into body hatred. I would then
forget my own words and praise her for doing something
'perfectly', at which she would correct me, saying it couldn't

be perfect: no such thing. I tried to promote her self-esteem by saying, 'Goodnight, beautiful girl,' every night; and then I stopped, fearing that she would equate beauty with self-worth. I watched her vigilantly for signs of wariness around cake. 'It's fine to have a slice. Or two if you are hungry.' No food, I insisted, was absolutely healthy or unhealthy, not even salad. I talked a lot about how it was good to be normal-sized rather than skinny or chubby, but a bit of chubbiness was fine too, especially in teenagers. I showed her exposés of how models in magazines are Photoshopped so that she wouldn't be taken in by those deceitfully lovely images.

So far (she is twelve), no eating disorder. But it's unlikely to have been all my little preventive measures that have saved her (if, indeed, they have). It's an extremely good thing for families to promote sanity around bodies and food, as far as they can, but in the end our attempts to save our children from anorexia are pieces of magical thinking, like throwing salt over your shoulder to blind the devil. There is no charm that can absolutely ward off anorexia. The current evidence on this puzzling and eerie disease is that the causes are more biological than social. While there is no single anorexic gene, up to 85 per cent of the risk of developing it is genetic.[20]

There has been a real sea change in clinical thinking about anorexia over the past two decades. The prevailing view among those who treat it now is that it is a largely heritable condition of the brain rather than a symptom of having an overbearing mother or seeing too many adverts featuring thin models. Scientists have identified a cluster of anorexic genes that have to do with the drive for perfection, the need for control and low self-esteem. Research in 2013 by a team of Cambridge scientists led by Simon Baron-Cohen found that adolescent girls with anorexia showed elevated autistic traits on cognitive tests, compared to a control group.[21] The suggestion is that the characteristic brain structure of anorexic patients (the neural phenotype) is strongly inclined towards systems, like the brains of autistic children. Many studies on anorexia have shown that sufferers display a high degree of social anxiety and difficulty interacting with others.[22] Both autism and anorexia are associated with social anhedonia: an inability to find pleasure in many of the social interactions that others find enjoyable. The theory is not that anorexia and autism are the same – or that every anorexia sufferer is socially withdrawn – but that they share certain neural traits that are expressed in different ways. It is striking that whereas the male to female ratio is roughly 10:1 for autism, with anorexia the situation is reversed, with a male to female ratio of 1:9. Baron-Cohen noted that the rigid mental attitudes of anorexia mirrored the narrow and repetitive behaviour of autism 'but in anorexia happen to focus on food or weight'.[23]

Patients with anorexia have brains that work slightly differently from the rest of the population, although whether the brain dysfunction is cause or consequence of starvation is not so clear. Neuroimaging has revealed various forms of cognitive impairment in anorexic patients. In particular, anorexics have a poorly functioning insula, a part of the brain that helps regulate anxiety. The insula is also crucial for flavour recognition. Some of this brain malfunction may be a response to a lack of food. But it seems that the insula of anorexics is still impaired after recovery, suggesting a structural flaw that predates the onset of the illness. One study measured the brain response of sixteen recovered anorexics to the pleasant taste of drinking sugar water. In contrast to a control group, these women had reduced activity in the insula when they drank the sugar water. It was as if their brains had difficulty recognizing pleasure.[24]

As with any genetic inheritance, however, having an 'anorexic brain' is not enough to give you an eating disorder. You might have anorexic genes but never get ill. Carrie Arnold, who is both a recovering anorexic and a biologist, describes the condition as arising from 'a complex interaction between malfunctioning hunger signals, anxiety, depression, and difficulties with decision making'.[25] If the causes of anorexia are more biological than social, this could be taken as good news for parents. It absolves the families of anorexic children from the crushing sense of guilt that affects so many. Arnold notes that her own parents were relaxed around food and never counted calories or pressurized her to lose weight. In most cases – though there are exceptions, where eating disorders are triggered by abuse or cruelty – parents are not 'to blame' after all, except insofar as they have passed on their genes. A family history of anxiety or depression places children at a markedly higher risk of developing an eating disorder. The downside to this is that if parents are not primarily to blame, there may be little a parent can do to prevent their child from becoming ill.

One of the many scary things about anorexia is how young the sufferers sometimes are. A 2011 survey of eating disorders in Britain found that, while the incidence of the illnesses was stable overall, it was increasing in younger children.[26] Out of all the new cases of children with eating disorders, 59 per cent – whether boys or girls – were preadolescent. It was not uncommon for them to be as young as ten or eleven. Some – though this was far more unusual – were eight; or seven; or six. It scarcely seems possible that a child so young could have the distorted body image and fear of fatness of anorexia. Part of the magic of childhood for most of us, looking back, was the sensation of freedom in your own body – the feeling that these legs were made for skipping. How cruel that any seven-year-olds, who should be eating ice lollies in the park without a care in the world, should be calculatedly starving themselves.

The obvious explanation for anorexia arising in children so young is that something has gone hideously wrong in our culture. There's no doubt that anorexia and bulimia are most common in Western or Westernized societies that revere thinness while pushing foods that make it very difficult to be thin. Anorexia usually starts with a period of dieting. Maybe a child decides to cut out dessert because she has been told at school that sugar is unhealthy or maybe because she has been teased for the way she looks in a swimming costume. The ideal female body type depicted in advertisements in glossy magazines is estimated to be possessed by just 5 per cent of women, leaving the other 95 per cent feeling potentially unworthy. As for boys, the ideal superhero body – quads like tree trunks, a small waist and an ability to fly around sky-scrapers – belongs to precisely no one. Little kids hear their parents talking about how they wish they could lose weight or calling pudding 'naughty' or using 'skinny' as a compli-ment; and anorexia would seem to be one logical response.

But eating disorders are not just a creation of modern life. If our culture causes anorexia, how could there have been documented cases of anorexia as long ago as the 1890s? In 1895 a doctor at an English children's hospital described 'A Fatal Case of Anorexia Nervosa' in an eleven-year-old girl. 'She had a wild, hysterical appearance, was very restless, and refused all food,' recorded the doctor.[27] He tried to feed her up with beef tea, brandy and milk, but after fifteen days in the hospital, she got a fever and died.

One hundred years later, in the mid-1990s, a seven-year-old girl, VE, was admitted to Massachusetts General Hospital. She weighed just 57lb (26kg) but stated to doctors that other children would 'like her more' if only she could get down to 50lb. She no longer ate regular meals and drank nothing but water. She was fearful of eating, or even of chewing her own fingernails, for fear of gaining weight. She spoke in an 'infantile manner' yet would also make comments about her thighs and stomach being too fat. She was under the impression that folds of fat hung down over her pyjamas, when in fact her tiny body was 'engulfed' by her clothes. Before her hospitalization, VE was involved in competitive dancing, figure skating and gymnastics. Her mother had wanted to be a dancer herself and said she could imagine VE dancing on Broadway. There were conflicts in her parents' marriage. VE's mother had a tendency to become enraged whereas her father would withdraw and leave the room rather than argue.

At first glance, this sad case looks like strong confirmation for the view that parents and culture cause anorexia. It's a lot of pressure for a seven-year-old, to be engaged in not one but three highly competitive individual activities – skating, dancing and gymnastics – all of which place a premium on being thin. Four months before VE was hospitalized, her mother had discouraged her request to give up dance lessons.

But the assumption that ballet and elite sports 'cause' anorexia has been questioned. Prima facie, children who do physical activity for several hours every day already look pretty similar to sufferers of 'anorexia athletica', who engage in compulsive exercise. Disciplines that emphasize leanness have a higher prevalence of eating disorders than endurance or ball sports. In one study, more than 80 per cent of female ballet dancers were estimated to have an eating disorder over a lifetime, yet other studies suggest that the prevalence is lower than 10 per cent.[28] Recently, there's been a rethink on the role of physical activity in eating disorders. In old-style treatment, anorexics were discouraged from doing exercise, in case it aggravated their illness. However, a major review of medical databases in 2013 found that supervised exercise could actually aid the recovery of anorexics, by building strength and cardiovascular fitness and alleviating the symptoms of depression.[29]

In hospital it became clear that VE's perfectionist tendencies predated and went beyond her recent involvement in dance, skating and gymnastics. From her earliest years, her parents had found her to be 'difficult' and highly competitive with her peers. She seemed to greatly fear any sign of weakness in herself and was distraught about some recent poor maths scores at school. In hospital, she worked hard to show she was a good patient and was needy for praise from staff. Anorexia sufferers often say that, long before they started attempting to lose weight, they remember feeling anxious, fearful, socially vulnerable and obsessive in various ways. Around two-thirds of people with anorexia also suffer from an anxiety disorder. While neither of VE's parents had an eating disorder, both suffered from bouts of depression. Her mother had twice been hospitalized with postnatal depression and she had been treated for OCD.

Without all the pressure surrounding skating and gym competitions, VE's anorexia might have taken longer to manifest itself. In addition to refeeding her, VE's cure involved switching her hobbies to team sports and group activities such as soccer and Girl Scouts, which allowed her to 'feel special'. But her anorexia was not 'about' the sport or the dance. Plenty of people do manage to become elite athletes or top dancers without developing eating disorders. With the family history of depression and OCD, VE had a biological make-up that would have made her vulnerable to eating disorders even without the figure skating, dancing and gymnastics.

Anorexia tends to occur when someone who is genetically predisposed to the illness suffers some kind of stress or trauma. But often the trauma is nothing more or less than puberty. The natural weight gain that occurs as children's bodies change into adult ones can prompt body dissatisfaction: are these strange swelling limbs really mine? Anorexia may be a way for girls to desexualize themselves and return to the safety of prepubescence: as the weight falls, breasts and hips melt away and periods stop. The hormones of puberty also seem to play a role in triggering anorexia in some people. New data from twin studies indicate that estradiol, the female sex hormone, can 'switch on' the genes that predispose some individuals to anorexia.[30]

With the falling age of puberty, it follows that the age at which children develop eating disorders is also dropping. Susan Ringwood is the director of Beat, the leading eating disorders charity in Britain. Ringwood confirms that rising numbers of younger children are calling their helpline. 'We're not quite sure what is doing it,' she says. One possibility, however, is the younger onset of puberty. 'The average age of puberty has dropped by about five years in the last fifty years,' notes Ringwood. Given that we know that the onset of puberty increases the risk of developing anorexia, it would be surprising if there were not some link between this much earlier puberty and the early anorexia.

If this is correct, then some of this current childhood anorexia, paradoxically, has its roots in the obesity crisis. The causes of puberty are not easy to unravel but there does seem to be a clear case that in girls, greater BMI is associated with earlier onset of periods and breasts. 'It's primarily driven by weight,' says Ringwood. 'Forty-two kilos and you are in.'[31] In 2000 it was found that one in six girls in Britain were showing signs of puberty as early as eight. One in fourteen boys aged eight had pubic hair, compared to one in a hundred and fifty for their father's generation. 'We know that the biological mind of puberty starts to develop around two years before the physical effects,' says Ringwood. There is a domino effect: from child obesity to premature puberty; from premature puberty to eight-year-old anorexics. 'It's the double whammy,' says Susan Ringwood. 'You are starting to develop an adult body when you have even less of an adult mind.' When young children become gripped by anorexia, the illness seems to escalate faster than it does with teenagers. A comparison of child and adolescent anorexics found that the children lost weight faster and were likely to have a lower percentage of their ideal body weight at the time they – or their parents – sought medical help. This was all the more worrying, given that they were still of an age where they needed the best nutrition to help them grow and develop long-term bone density.

The only good thing about anorexia hitting the very young – and admittedly it's not much of a silver lining – is that they tend to have better rates of recovery and a shorter duration of illness than older sufferers.[32] In some ways, being a child helps the situation. 'If they are to recover,' says Susan Ringwood, 'children with anorexia do need to be made to eat.' The advantage of being a child is that you are already in the habit of having other people feeding you. You are also in the habit of listening to adults tell you what to do, so it's not so strange when they tell you there is simply no option but to eat. In recovery, this childish obedience can be very helpful, for as long as it lasts.

When a child has an eating disorder, family meals can be miserable, soul-sapping occasions where children lie and parents wheedle and very little is eaten by anybody. Or they can be exercises in make-believe, where everyone politely pretends not to notice that one person has eaten no more than a couple of pieces of cucumber and half a yoghurt.

But family dinner is also a child's best hope of getting better. Viewed through the prism of anorexia, you see just what a powerful and therapeutic thing a meal can be. When it goes right, a child is, all at once, being given nutrition and love and a way to escape their prison of misery. It isn't easy to reach this point, for any of the parties concerned. A mother of a daughter who was severely anorexic for nine years described her frustration at reading over-optimistic accounts of 'parents who just insisted that the child ate, the child did, and roses grew around the door and life was lovely again'.[33]

With anorexia and related eating disorders, the stakes are even higher than with selective eating. Restrictive eaters are not actively trying to starve themselves; anorexics are. One of the horrible truths about anorexia is that not to recover so often means death. A systematic review of the literature in 2002 looked at studies involving more than 5,000 anorexics. It found that while there was a 'good outcome' for around half of anorexia sufferers (meaning that all symptoms had gone) and a 'fair outcome' for around 30 per cent (meaning improvement with some residual symptoms), there was a 'poor outcome' for over 20 per cent, meaning that the disorder was chronic. With anorexia, a poor outcome means death in some cases.[34]

At the Maudsley Hospital in south London in the 1980s, however, some therapists discovered that they could achieve much more hopeful outcomes with anorexic patients than the statistical norm. They did this by focusing more intensively on the symptoms of the disorder: the eating itself. Therapists noticed that when nurses sat with the patients as they ate, talking to them and sometimes rubbing their backs, they could create an atmosphere of such kind persistence that it was 'impossible . . . not to eat'.[35] These Maudsley doctors had the shrewd thought that perhaps parents could be taught to perform the same role at home. This was the kernel of the current movement of 'family based treatment' (FBT), sometimes called 'the Maudsley approach', although most of the research on which it is based was done in America at Stanford and the University of Chicago in the 1990s. FBT works on the basis that a child with anorexia needs to be systematically 'refed' by her parents, until she is well enough to take responsibility for her own eating again. Research by Daniel Le Grange and James Lock, two of the leading lights of FBT, suggests that for anorexic patients who are younger than eighteen and with a relatively short duration of illness, recovery rates can be as good as 90 per cent, with full remission after a year and again after five years. FBT – when followed very systematically – achieves these remarkable recovery rates by doing exactly what most eating disorder therapists had been trained not to do: allowing the parent to take control of the child's eating.

Traditional eating disorder treatment was based on the idea that parents were to blame. *The Golden Cage* by Hilde

Bruch (1978), a German psychoanalyst working in America, was an influential book that described the parents – especially the mothers – of anorexic girls as monsters who stifled their children with impossibly high expectations and an atmosphere of neurosis.[36] Bruch felt that in order to recover, a patient needed to separate from the family. Individual therapy would encourage a patient 'towards independence'. There was no question of family meals being used as part of the treatment, because family meals were seen as the cause of the anorexia in the first place. On Bruch's model, parents were often warned that they must not sit and eat with their children, lest their presence was oppressive. They should offer no judgement on what their child ate, but allow them to decide their own eating. In some cases, therapists felt that a 'parentectomy' was advisable: a total separation of parent and child. The thinking was that since anorexia was not really 'about' the food, the child would choose to eat once she had worked through her other issues. But the point about having an eating disorder is that the child is *not* in control of her own eating. Left to her own devices, she will, likely as not, revert to the disordered behaviour, whether it is bingeing or starving. With the traditional treatment, anorexia clinics found that their patients might recover in hospital, where they were being actively fed – either by tube or with real food – and then relapse very quickly when they got home. Which was hardly surprising given that the parents were being told not to interfere in their child's eating.

FBT turns this dynamic around. It is predicated on a non-judgemental attitude towards parents. This is not to say that a family dynamic never contributes to an eating disorder, but that what a very sick child needs is urgent treatment rather than endless discussions of what made them ill. Guilt is a crippling emotion that makes parents feel hopeless and unable to act. The idea of FBT is that parents must feel in charge of getting their child to eat again and must therefore stop blaming themselves. Once they forgive themselves, they are in a position – with the help of a therapist – to start the hard task of refeeding, which is a little bit like teaching an infant how to eat solid food again. Again, the needs of the eating disorder sufferer are like a warped version of the travails we all face when learning to eat.

As with weaning, 'refeeding' is a slow process, requiring stamina. To start with, a parent may be happy if the child manages a meal of mashed pumpkin with a teaspoon (many anorexics revert to using baby cutlery). As time goes on, you expect more of them, upping the calories in stages. They need to add new foods to their repertoire, like the selective eaters. You refuse to give them low-fat options. No meals are to be skipped and the child is encouraged to take one more bite than they wanted to. The food is never forced, but nor is the child allowed to say they don't feel like eating. James Lock argues that it is a mistake to respect the voice of the child when they say they do not want to eat, because that is the illness talking.[37]

At the start of the FBT process, the family will have one or more 'coached meals' at which a therapist counsels parents in how to manage family food, such that a child will eat. By the time they arrive at an eating disorders clinic, families often say they have 'tried everything' at mealtimes, but the odds are that – as with parents of selective eaters – they haven't felt able to pursue any of the techniques consistently. Many of the families seen by James Lock do not have regular mealtimes, just grazing on food at ad hoc moments. The whole family, not just the patient, needs to relearn how to eat breakfast, lunch and dinner, with structured snacks in between. Siblings, too, must be included, though parents are trained not to fall into the trap of comparing what different people around the table are eating. The 'coached meal' teaches parents how to stop treading on eggshells around their child's eating: to sit close and repeat calmly and assertively that she must eat the meal in front of her, even if she refuses or cries or says she hates you. The parents need to agree between them before the meal starts how much they expect the child to eat and what the consequences will be if they don't (no computer games for a day, for example). With divorced parents, Lock goes so far as to say that the child should live for the time being with the parent who is better able to manage the meals.

In her memoir of living with her then fourteen-year-old anorexic daughter Kitty, *Brave Girl Eating*, Harriet Brown, an advocate for FBT, describes how a refeeding meal might go.[38] Brown lays out Kitty's breakfast of a bowl of cereal with milk and strawberries. Kitty says she wanted cottage cheese instead. Brown says there is none. Kitty complains the cereal is soggy. Brown makes a fresh bowl but then insists as calmly as she can that Kitty 'sit down and start eating'. This whole rigmarole goes on many times a day, with Kitty often sobbing that the food will make her fat and Brown protesting that food is her 'medicine' and she must eat. Brown or her husband take it in turns to stay with her for an hour after each meal to prevent the risk that she will run to the bathroom and purge. There are many such meals – Kitty needs to eat a snack every couple of hours. Four years on, Kitty has recovered to the point where her parents feel it is safe for her to go off to college by herself and take responsibility for her own meals. There are still relapses when the 'demon' returns to the dinner table and Kitty's weight drops, but at least they all feel they have done everything they can to normalize her relationship with food. More importantly, Kitty herself now has an approach to food she can follow to turn her situation around when she starts to slip back into losing weight. Food is medicine.

One of the many hard aspects of refeeding is that it isn't enough for the patient to eat an amount of food that would be healthy for a normal-weight person. Someone with anorexia needs vastly more calories than before, to regain the weight needed for body and brain to recover. Anorexics would never 'choose' of their own accord to drink a 1,000-calorie milkshake, but after they recover, patients often say that it was strangely liberating to have their parents telling them that they had no option but to eat, because it reduced the shame. Families need to become experts in which foods offer the most calories without filling up the child to the point where their stomachs hurt. It is the opposite of the way most of us try to eat, seeking out the maximum food for the minimum calories.

This refeeding process may be even harder for those anorexics who develop the disease later in life and are without parents at hand to help them. Some years ago, I wrote an article on women who were battling anorexia in their thirties, forties and fifties.[39] Among those I met was Jane, a reticent fifty-three-year-old teaching assistant, who described the humiliation of being a middle-aged anorexic. For her, the misery of anorexia was compounded by a sense of shame, that at her age she should have 'known better' as she put it. At her lowest point, Jane lost 5 stone from an already slim body. Once, she felt so despairing, she took a hammer and smashed her own hand. She was put in a therapy group with six 'trendy' teenage girls and expected to open up about how she felt. How Jane felt was: 'Why should I share my innermost feelings with a group of strangers?' Another obstacle to Jane's recovery – in common with the other older anorexics I interviewed – was that she was the one in the family who provided for everyone else. Jane was very good at feeding others, but feeding herself was another matter. She prepared lavish, ambitious meals for her husband and two sons, while she nibbled on an apple or a yoghurt. On the rare occasions that she ate out with her husband, she could be reduced to tears by the arrival of a bowl of soup. When I met her, Jane was slowly teaching herself to eat again, and had managed to edge up to 1,000 calories a day: not enough – she was still painfully thin – but just sufficient to keep her out of hospital.

For some adult anorexics, the best course of treatment may be a residential programme where the patients – of whatever age – can enact the role of children in the protective setting of a family meal once again. I visited Newmarket House in Norwich, a specialist treatment centre for anorexia that felt more like a spacious home than a clinic, with colourful sofas, and appetizing cooking smells in the air. I met Beth, who was in her thirties, a mother of four. Like Jane, Beth was a confident cook, and took great pride in the birthday cakes she baked for her children, but she struggled to allow herself to eat anything but lettuce and tomatoes. She wished she could disappear, she said, and was still far from fully recovered. But at least the structured meals of Newmarket House – at which the nurses feel more like family members than therapists – gave Beth an environment where others took care of her eating for a change.

With some eating disorders, however, being older and more independent seems to be beneficial for recovery. Bulimia tends to strike at a later age than anorexia (in a review of 5,653 cases of bulimia, the average age of onset was seventeen, but often it starts in the twenties).[40] A study of forty women who had fully recovered from bulimia found that they tended to be self-motivated about getting better and did not like the view that 'one is powerless over one's problems'.[41] Eighty per cent of these recovered bulimics had ultimately been motivated to change by their own desire for a better life and weariness at the symptoms (when they were ill, they vomited on average twenty-two times a week). Although most of them benefited from professional help, nearly half of them backed this up by reading self-help books. Another study found that among a group of bulimics in Austria, more of the patients became symptom-free by using guided self-help, working through a manual by themselves, than by being given a course of CBT.[42]

Learning a new balanced way of eating after bulimia or binge eating is very different from refeeding an anorexic. Instead of boosting calories, a bulimic needs to find a reliable way to limit each day's food, avoiding anything that might trigger an episode of bingeing. Unlike a selective eater, a bulimic needs to teach themselves to become *less* omnivorous. One forty-five-year-old recovered bulimic described the strict regime she had created for herself, as a result of which she had celebrated eighteen months without any symptoms.[43] She shopped in very small quantities to make bingeing impossible, and ate five small meals a day of fish, meat, fruit and vegetables. She breakfasted on tinned tuna or cold chicken, because bread would remind her too much of bingeing and create a temptation to purge. Wheat and dairy were now eliminated from her diet. For an anorexic, such rigid food rules might be a dangerous path to take, but for a bulimic, there can be liberation in limits.

There is at least one respect, however, in which the situation of anorexics and bulimics is very similar. Before addressing what to eat, the most urgent matter is how to eat. Phase one of recovery from bulimia is the reintroduction of regular mealtimes: no bingeing, no starving. Slowly, the days regain a sense of rhythm. As anyone who has ever suffered jet-lag knows, few things are more disorientating than a warped sense of time. Some of what makes bulimia nightmarish – in common with other eating disorders – is that it disrupts the daily tempo of meals. One recovered binge eater spoke of how she used to live in a 'food-fuelled haze' but had now found that by allowing herself regular clearly defined meals, she had regained a feeling of certainty.[44] Lunch becomes a meaningless concept when you have already eaten – and possibly purged – a whole box of cereal by mid-morning. When you are eating all the time, food curiously loses much of its joy, along with its sense of ceremony and sociability.

Once again, the experiences of eating disorder sufferers are on a continuum with the rest of us. It is hard for anyone to live well when meals are not given the attention they are due. As the New Yorker writer Adam Gopnik has asserted, 'the table comes first', meaning that before we can resolve our endless quandaries about food – such as 'where the zucchini came from and how far it had to travel' – we should first establish the basic paradigm that at certain times, every day, we stop, we sit and we eat.[45]

In many ways, the needs of a bulimic or anorexic are not so different from those of an adult picky eater like Diane or just an averagely screwed-up human being with a desire to lose weight. An individual with anorexia 'is disconnected from her internal experiences' and cannot read her internal hunger signals effectively, notes one scholarly article on anorexia.[46] But most of the population, as we have seen, is similarly disconnected from internal signals about when, what and how much to eat. The difficulty is that those whose disordered eating is less extreme are likely to have less help: when you sit down to eat, you are both the parent and the child, the doctor and the patient. Like the anorexic patient faced with the 1,000-calorie milkshake, many of us would never 'choose' to eat a plate of healthy food over a fast-food meal, but if we can give our bodies the food it needs often enough, in a kind and persistent enough way, we may eventually start to recover. Eating is about the food. What all of us need is to find a way to eat regular meals, to take pleasure in a variety of foods, and to be able to eat them without being consumed by negative emotions.

It is startling to hear the message, from eating disorder therapists, that nourishing, health-giving family meals, eaten in loving company, are so important for a child's wellbeing that everything else in life must be made secondary to them. Most families – most people – do not live like this. Gone – thankfully – are the days of a patriarch ruling the family from the head of the table. The breakdown of strict table manners – children should be seen and not heard! – has been emancipating in some ways. But as a society, we haven't quite figured out what a new structure for meals would look like that isn't just a hasty sandwich in the car on the way to something more important. The experience of eating each day around a table is given second billing to other activities: homework, after-school pursuits, Instagram and email. In a busy life, the organization required for regular shared dinner can seem unattainable, and even if you can manage the shopping and the cooking, parents often hesitate to assume the authority to gather everyone together to eat, never mind to insist that everyone eats the same food. But the experience of eating disorders shows that this is partly a question of priorities. When eating becomes life or death, and each new bite is a celebration, you may discover that none of the other stuff was quite as important as sitting and breaking bread together.

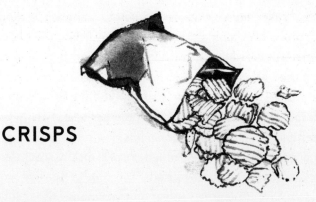

CRISPS

I knew a family whose children collected – and then
ate – crisps from different countries. Whenever friends
went abroad, they asked them to bring a packet or two back.
They had eaten curry chips from Belgium and shrimp crisps
from Thailand; crinkle-cut from Australia and paprika ones
from Germany shaped like kangaroos. Their whole idea of
global food was variations on fried potatoes.

People with selective eating disorders often find that
potato crisps (usually plain salted ones) are one of the 'safe'
foods they can best tolerate. 'Bob K', one of the founders of a
support group for adult picky eaters, says that crisps satisfy his
two main requirements from food: they are plain tasting, but
crisp and crunchy in texture. They are also reassuringly beige
in appearance. In 2012 a fifty-four-year-old woman dubbed
by the newspapers 'the world's fussiest eater' told reporters
she ate only three foods: milk, white bread and fried potatoes,
whether in the form of fries or crisps. Of these, the crisps
were her favourite because they were 'so salty and fresh and
potato-y'.

Selective eaters are not alone in this love of crisps. It's
tempting to feel that most of the planet is on the feeding
disorder spectrum, judging from our crisp habit. Some stack

them up like Jenga blocks, to cram in as many as the mouth will hold; others nibble them one by one, licking off the salt before biting. Fried slivers of potato were once an aristocratic treat, used as a garnish to roast game birds and eaten in small quantities. Those days have gone. In 1964 the British ate an average of 250g of crisps per person per year; twenty years later, this had jumped to 1.33kg. It is now more than 3kg, not allowing for all the extruded crisp-like salty snacks and crackers that are also devoured.

How have we learned to eat quite so many crisps? John S. Allen, author of *The Omnivorous Mind*, notes that crispiness is a nearly universally loved texture across different cultures. Part of the appeal is that chewing crispy foods activates our sense of hearing as well as smell and taste. Making that loud crunching sound is part of the pleasure: it staves off boredom and makes you eat more. Allen suggests that our penchant for crispy food may go back to our primate ancestors, for whom crunchy insects were a valuable source of protein.

But like so much about the way we eat, our instinct for crispiness has outlived its usefulness. Almost all the commercially produced crispy foods – from crisps to fried chicken to breadcrumbed nuggets – are ones we'd do well to eat less of. I can't deny that salty fried food can be delicious. One way forward could be to get your hit of crispiness in vegetable form, as beetroot and parsnip crisps, say. It is possible to make vegetable fritters so inviting – cauliflower pakoras, aubergine tempura, sweetcorn pancakes – that crisps seem dull by comparison. But it is still hard to replicate the crunch.

CHAPTER 8

Change

The child does not necessarily even notice that he
has learned, only the adult notices.

Report on the Sapere system
of food education, Finland

From the perspective of almost everyone else in the world, the
Japanese have an enviable relationship with food. Japanese
cuisine – with its focus on fresh vegetables, even fresher fish,
delicate soups and exquisitely presented rice dishes – has a
global reputation for healthiness. Japan has somehow managed to achieve the ideal attitude to eating: an obsession with
culinary pleasure that is actually conducive to health. The
Japanese must be doing something right in the way they eat,
given that they live longer on average than people from any
other nation.

There is a higher concentration of Michelin-starred
restaurants in Tokyo than in Paris or New York or London.
In Japan, food filters into every aspect of the culture. There
are theme parks devoted to sushi and songs are sung to

noodles ('The slippery saltiness cries out/Is it my tears or maybe a dream?').[1] Yet at the same time, for a rich nation, Japan has remarkably little obesity. Admittedly, far more people – especially men – are obese than twenty years ago and Japanese adolescents eat more junk food and have more eating disorders than the previous generation. But systematic figures from 2013 suggested that just 3.3 per cent of Japanese women were obese, compared with 20.9 per cent of women in Poland, 33.9 per cent of women in the US, and 48.4 per cent of women in Egypt.[2] One of the factors keeping Japanese weight in check is a controversial law, introduced in 2008, under which companies can be fined if too many of their employees exceed a maximum waistline (33.5 inches for men, 35.4 inches for women).[3] The very fact that the Tokyo government could succeed in passing such a law, however, is a sign of how far Japanese eating habits were already under control. Almost the only places in the world that have lower obesity averages than Japan are countries such as Ethiopia and North Korea where there is widespread hunger and food itself is scarce. Japan is almost the only country that has such low obesity without a starving population.

It's easy to look at Japan and think that there must be something essential in the culture that makes the nation eat so well. Eating dainty and beautifully presented meals seems of a piece with origami, Buddhist temples, silk kimonos and cherry blossom. In China many women regard eating 'Japanese food' – meaning rice, vegetables and miso soup – as the secret to health and beauty. There are signs that the Japanese themselves consider their excellent cuisine as an essential part of what it means to be Japanese. The Japanese Ministry of Agriculture, Forestry and Fisheries sells the idea that Japanese cuisine has always been envied the world over.[4]

For those of us who don't live in Japan, the cult of Japanese food can feel dispiriting. How easy, we think, it would be to eat healthily if only we were in Tokyo! Perhaps we too might breakfast on miso soup and fish and dine on greens, rice and tofu. What tiny waists we would have, and such healthy hearts. Our happiest childhood memories would be of our mother's soba noodles and seaweed, rather than cereal milk and junk foods. We would find a way to square the circle of enjoying food without overindulging. But since we are not in Japan, the thinking goes, we are probably doomed to eat badly. We could never eat like they do in Osaka or Tokyo. How can you eat like Japan without being Japanese?

This path of thinking misses the fact that the Japanese themselves have only been eating the way they do for a very short time. We are often fatalistic about our patterns of eating, denying our capacity for change, and this fatalism can be seen both in individuals and at a wider social and cultural level. With our own personal diets, we often convince ourselves that there is something vital within us that prevents us from ever eating differently. Meanwhile, we are also fatalistic about the diets of whole populations, assuming that once an unhealthy 'Western diet', high in refined carbohydrates, has been adopted, there is no going back. We presume that something as huge and all-encompassing as a food environment is not subject to modification. Indeed, when even quite modest efforts at reform to the food system are made – such as Mayor Bloomberg's failed law to limit the size of sodas sold in New York City – they are attacked as revolutionary. There is a deep resistance to the idea of dietary change, at both a cultural and an individual level. And yet, once you accept the premise that eating is a learned behaviour, it follows that changing eating habits must be – if not likely and certainly not easy – at least possible.

Actually Japan itself is a model for how whole food environments can change in positive and unexpected ways. Until the twentieth century, Japanese cuisine had a reputation far inferior to that of China. It is telling that while Japan borrowed many aspects of eating from China – including noodles and chopsticks – China never chose to copy Japan until the late twentieth century.[5] Food in Japan was neither varied nor appealing and there was never enough of it. From the seventh century AD to the twentieth century most of the Japanese population was in a state of hunger and gastronomic isolation. Dinner was seen as a necessary fuel rather than a pleasure, let alone an art form. Unlike their neighbours in Korea, the Japanese had no love of spice. While their counterparts in China composed food poems and cookbooks and relished the social aspect of meals, Japanese diners sat mute at the table. During the Tokugawa era (1603–1868), when Japan was largely sealed off from the outside world by a policy of national seclusion, Japanese visitors to China were shocked by the Chinese habit of conversing while eating. As late as the 1930s, it was the custom at family meals in Japan to eat in silence, while consuming some fairly basic rations of rice and pickles.

Barak Kushner, a distinguished historian of Japan based at the University of Cambridge, suggests that until recently Japanese cooking was just 'not very good'.[6] The fundamental techniques of stewing and stir-frying were only adopted as late as the 1920s.[7] The traditional diet was low on protein, often dangerously so. Kushner notes that until the twentieth century, the Japanese ate far less fresh fish than we might expect (for middle-income families, it was a weekly rather than a daily food). For centuries, a typical meal in Japan would be some rough grains accompanied by something like shredded yam leaves and daikon radish, with miso and pickles: not a terrible way to eat, but not a very joyous or varied one either.

I first met Kushner at a 'ramen noodle workshop' he gave at a little noodle place in London's Soho. Kushner told me that when he first arrived in Japan as an English-language teacher, in the 1990s, much of the cuisine repulsed him, particularly the raw fish. He grew up in New Jersey thinking that a certain kind of chocolate cake called Ring Dings, in little foil wrappers, were the pinnacle of deliciousness. But since he knew he would be in Japan for a while, and he was hungry, he kept trying the cuisine. Now, twenty years on, he is married to a Japanese woman and says he would rather eat Japanese food than any other kind.

At Kushner's workshop the assembled company of food writers devoured steaming bowlfuls of freshly made ramen: a salty 'shio' broth of pork stock, seasoned with seafood extract, topped with a mound of springy, slippery noodles, a slice of tasty pork, half a soft-boiled egg and some dark greens. Kushner ate with gusto and instructed the rest of us how to slurp, drawing in air to cool down each mouthful. 'This is not a slurp for speed; this is a slurp for enjoyment,' he remarked.

This fine dish – ramen noodles – has become one of Japan's many foodie obsessions. It is far from the packaged instant kind known as a cheap staple in the lives of students. Though it is humble and inexpensive compared to sushi, 'good ramen is difficult and time-consuming to make', as Kushner writes in *Slurp!*, his superb history of ramen in Japan. The broth – which varies in the different regions of Japan – needs to be carefully simmered; the noodles are cooked fresh for each order; the flavourings on top are arranged with artistry and care.

Kushner's real subject is not ramen but the way a country can completely change its diet and attitudes to food. 'Japanese cuisine is neither timeless nor unchanging,' he argues. A few weeks after the workshop, I met Kushner for tea. He told me how much he disliked 'essentialism' about food: the notion that there is some inherent 'Japanese-ness' about Japan that makes people eat a certain way. Many of the dishes most loved in Japan today were borrowed from China and Korea. Kushner's research has taught him that the true story about the current Japanese diet is that it came about due to 'a multitude of factors' including travel, industry, politics, geography, war, the rise of cities; and science. The concept of 'delicious' was born in Japan in 1908 when a chemist called Ikeda discovered a 'fifth taste' called umami that was neither bitter nor salty nor sweet nor sour but something more wonderful and compelling than any of these. Umami is the savoury meatiness in seaweed and miso and soy sauce. It is, to a large extent, the concept that enables Japanese cuisine to be healthy and attractive at the same time. In the West the word 'delicious' is likely to conjure up something laced with sugar, fat and salt, whereas in Japan it signifies a flavour found in mushrooms, grilled fish and light broths.

Yet it took Japan a very long time to reach delicious. Ramen, with its delicate balance of flavours and textures, goes against most of what the Japanese traditionally believed about food. For centuries, the wheat that goes into ramen noodles was seen as an alien grain. Supposedly, a meal was only 'Japanese' if it contained rice, though most people were forced to bulk out their rice with coarser grains such as millet and barley or, in really hungry times, with ground-up acorns. Noodles arrived in Japan with Buddhist monks from China in the Middle Ages, but until the twentieth century they tended to be made from buckwheat, or a mix of wheat and rice. The Japanese were also hostile to pork, which was seen as Chinese and slightly filthy. Yet over time, the Japanese came to enjoy noisily slurping up pork broth with wheat noodles so much that, as Kushner observes, 'contemporary Japan almost floats on a sea of noodle soup.'

Japanese cuisine did not change all at once but in stages. There were three crucial moments in Japanese history when new tastes were adopted, and each time the change happened as a matter of national urgency to improve the health of an undernourished population.

The first big changes to Japanese attitudes to food began during the Meiji restoration (1868–1912) when Japan became an empire and opened its borders to other nations for the first time. Finally, Japan started to make comparisons between its own diet and the ways that other countries ate. There were urgent discussions in the Meiji government about whether Japan's diet was making its people too weak and small to compete with the West. Educators argued that to be a true imperial race the Japanese must start eating meat and increase their consumption of milk.[8] In 1872 the emperor broke the 1,200-year taboo against red meat and informed his public that he was now a meat eater. It would be another fifty years before consumption of pork and beef significantly increased for most Japanese people. But the Meiji pro-meat propaganda did at least lay the groundwork for the idea that the Japanese did not have to eat the way they always had. For the first time, eating Western food could be a patriotic act. The opening up of the Meiji period planted the thought that people would nourish themselves better by abandoning their old food habits and learning new ones. 'We Japanese must open our eyes [to the benefits] of beef and milk' stated one piece of pro-meat advertising from 1871.

The second key period of change in the Japanese diet was the 1920s. The Japanese army was in a state of crisis. Many rural recruits were desperately malnourished on their traditional diet of miso, vegetables and grains. In 1921 a Military Diet Research Committee was set up to apply the latest in nutritional science to the army diet. Under Marumoto Shozo, the new director of military catering, what Japanese soldiers ate was transformed. Their meat ration was increased to 13kg of beef a year, a vast amount by Japanese standards. But the really remarkable change that Marumoto made was to switch the army diet to Chinese and Western dishes with a higher fat and protein content than the traditional foods. The reformed menus – which required new equipment for the mess kitchens – included pork cutlets, breaded chicken, noodles in curry sauce, beef stews, croquettes of various kinds and stir-fries. This was a bold move on Marumoto's part, and one that very few military caterers would have thought of. Soldiers, like footballers, are famously resistant to new foods. Yet in Japan the army recruits seem to have been sufficiently hungry to be grateful for these new exotic dishes, and by the end of the 1930s conscripts had developed a permanent liking for them. Meanwhile, the Japanese government extended the lessons of this new nourishing army diet to the rest of the population. Military cooks were ordered to give propaganda talks and demonstrations and make radio broadcasts, all trying to persuade Japanese mothers that by cooking army-style they could improve the strength of the nation.

But the Japanese only really started eating what we think of as Japanese food in the years after the Second World War. During the war, Japan suffered some of the worst hunger in any of the nations involved in the war: out of 1.74 million military deaths from 1941 to 1945, as many as a million were due to starvation.[9] The Japanese were reduced to acorns and rough grains and sparse amounts of rice, as they had been so often before. Japan was heavily dependent on imported food and was therefore hit especially hard when the war curtailed supplies. The ration rice – given in woefully inadequate quantities – became known as 'Five Colour Rice': white rice, stale yellow rice, dried green beans, coarse red grains and brown insects. Yet when the Japanese finally bounced back from hunger in the 1950s, they boomed to a state of unprecedented prosperity and a completely new openness to the pleasures of food.

The new Japanese adventurousness about food was partly a consequence of American postwar food aid. In 1947 the occupying US forces brought in a new school lunch programme to alleviate hunger among Japanese school children.[10] Before this, children would bring food from home: rice, a few pickles, maybe some bonito flakes but almost nothing in the way of protein. Many children suffered constant streaming noses from their inadequate diet.[11] The new official American lunches guaranteed that every child would have milk and a white bread roll (made from US wheat) plus a hot dish: often some kind of stew made from the remaining stockpiles of canned food from the Japanese army, spiced with curry powder. The generation of Japanese children reared on these eclectic lunches grew into adults who were open to unusual flavour combinations. In the 1950s national income doubled, people migrated from the land to tiny city apartments and everyone aspired to buy the 'three sacred treasures': a TV, a washing machine and a fridge. With new money came new ingredients: the national diet shifted from carbohydrate to protein. As the Japanese food historian Naomichi Ishige has explained, once levels of food consumption rose again to pre-war levels, 'it became clear that the Japanese were not returning to the dietary pattern of the past, but were rather in the process of creating new eating habits.'[12]

In 1955 the average person in Japan ate just 3.4 eggs and 1.1kg of meat a year, but 110.7kg of rice; by 1978, rice consumption had markedly decreased to 81kg per capita while people were now eating 14.9 eggs and 8.7kg of pork alone, not to mention beef, chicken and fish. But this wasn't just about Japan moving from privation to plenty.

More than anything else, it was a shift from dislike to like. Where once it was seen as extravagant in Japan to serve more than one or two dishes to accompany the evening's rice, now – thanks to the new affluence – it was becoming common to serve three or more dishes, plus rice, soup and pickles. Newspapers published recipe columns for the first time and, after centuries of silence at the table, the Japanese started to talk with great discernment about food.[13] They embraced foreign recipes, such as Korean barbecue, Western breaded prawns and Chinese stir-fries and made them so much their own that when foreigners came to Japan and tasted them, it seemed to be 'Japanese food'.[14] Perhaps thanks to all those years of culinary isolation, when Japanese cooks encountered new Western foods they did not adopt them wholesale, but adapted them to fit with traditional Japanese ideas about portion size and how a meal should be structured. When omelette was served, for example, it probably did not have fried potatoes on the side as it might in the West, but the old miso soup, vegetables and rice. At last, Japan had started eating the way we expect them to: choosily, pleasurably and healthily.

There was nothing inevitable or innate in the Japanese spirit that gave them this near-ideal diet. Instead of being dispirited by the way the Japanese eat, we should be encouraged by it. Japan shows the extent to which food habits evolve. We sometimes imagine that Italians are born loving pasta or that French babies have a native understanding of globe artichokes that runs in their blood. The food scholar Elizabeth Rozin has spoken of the 'flavour principles' that flow through national cuisine, often changing very little for centuries, such as 'onions, lard and paprika' in Hungary or 'peanuts, peppers and tomatoes' in West Africa. 'It would be as unlikely,' Rozin writes, 'for a Chinese person to season his noodles with sour cream and dill as it would be for a Swede to flavour his herring with soy sauce and gingerroot.'[15] Yet Japan shows that such unlikely things do happen. Flavour principles change. Diets change. And the people eating these diets also change.

It turns out that wherever we are from, people are capable of altering not just what they eat but what they want to eat and their behaviour when eating. It is startling that Japan, a country whose 'flavour principles' included little spice except ginger, should fall in love with katsu curry sauce made with cumin, garlic and chilli. A country where people once ate meals in silence has shifted to one where food is obsessively discussed and noodles are loudly slurped to increase the enjoyment. So perhaps the real question should be: if the Japanese can change, why can't we?

*

Just because dietary change can happen on a national stage does not make it easy to enact at a personal level. Imagine what it must be like to be one of the 3.3 per cent of the Japanese population who, despite the surrounding environment of slimness and good food, are in fact obese. The overweight in Japan are subjected to immense social pressure to change, including taunts of 'metabo' – short for metabolic syndrome – and unwanted belly-patting from strangers. And yet none of this is enough to induce weight loss. Personal dietary change cannot be forced in this way.

Slimming magazines, whose stock in trade is transformation, specialize in 'Before' and 'After' photos. The aim of these is to show the readers that losing weight is feasible, although the effect can be to make them feel even worse about not yet being one of the success stories. The 'Before' photo shows someone in oversized elasticated trousers, awkwardly avoiding the camera's gaze. The 'After' photo shows the same person half the size they were before, beaming in figure-hugging Lycra or a swimsuit. We are meant to be encouraged by the 'After' photo but the disparity between the two images is almost too great. When you are stuck in the 'Before' stage, 'After' can feel like another planet.

When I was an overweight seventeen-year-old, a couple of friends and I embarked – more than once – on diet and exercise programmes together. We would start with high hopes, but we tended to overdo things on the first few days, get worn out from hunger and aerobics and blow the diet before the end of the first week. It isn't easy to go jogging in public when you feel self-conscious and out of shape; it's no pleasure to 'snack' on celery when all you can think about is KitKats. One of the obstacles was that we didn't really believe we were the kind of people who could ever reach the 'After' stage. We used to complain among ourselves about the other girls, the ones for whom it seemed to come so naturally: the ones with the supermodel looks and honey-brown skin who genuinely preferred nibbling a small bowl of 'power muesli' and yoghurt to eating five slices of toast with peanut butter and jam. We found it deeply suspect that these girls would sometimes start eating something and stop halfway through because they said they just weren't hungry any more. They seemed as otherworldly to us as Japan. We could never be like them. We imagined that their ability to eat well was something ingrained and essential, impossible to imitate.

It is only now that I am through on the other side, in a place where food is not such a problem, that I can see our mistake. The healthy way these girls ate was not something they were born with, like hair colour. They were habits and preferences that they had somehow acquired through a combination of circumstances and upbringing, just as we had somehow learned to cheer ourselves up with toast, reward ourselves with sugar and finish everything on our plates even when we were stuffed. It is absolutely possible, I have discovered, to become the kind of person who craves a salad more than a sandwich – particularly if the salad is a tangle of tasty greens dressed with something flavoursome like anchovy, oil and lemon with buffalo mozzarella on the side. There have been encouraging developments in home cooking in recent years. As a reaction to the mainstream food culture, a new version of 'healthy eating' has emerged, one that – quinoa aside – lacks the austere mood of 1970s health food. This new vegetable-based cooking is wisely driven by flavour as much as nutrition: crispy sprouts, Vietnamese summer rolls stuffed with mint and peanuts, smoky roast red peppers, hearty chickpeas and garlic, sweet potato fritters. The food writer Diana Henry uses the phrase 'a change of appetite' to describe the appeal of this way of eating.[16]

A change of appetite does not involve a total change in personality. You still like the same music and films. You may never – despite what the health pages of magazines sometimes imply – develop supermodel looks and smooth honey-brown skin. But if you can find a way of actively desiring – and therefore eating – a wide variety of healthy foods, the odds are that you will feel better, have more energy to exercise, get ill less and enjoy meals more, because you eat without guilt. Considering that most of us consume well over a thousand meals every year, this is a very good place to reach.

The great question is how you get there; and how you don't. Anyone whose weight is edging upwards, for whatever reason, may find themselves on the receiving end of unsolicited advice. People around you, from colleagues to family to doctors, have opinions on how you could change: diets you should try, things you shouldn't eat, 'helpful' hints about where you are going wrong. They seem to be under the impression that their finger-wagging or hinting will make you change, as if you haven't noticed your own weight gain. But if it were possible to make other people change their eating habits through rational suggestions, then we would surely all be slender lentil-eaters. 'One would think,' observes a textbook on personal change, 'that the very real threats of kidney failure, blindness, and amputations would be sufficient to motivate people with diabetes to keep their blood glucose under control.'[17] In many cases, however, these terrifying threats are not enough. Neither fear nor advice is a good motivator of change. Having other people try to fix you is one of the things that paradoxically holds you back from reaching that magical place called 'After'.

Dympna Pearson talks in a gentle Irish brogue, so quiet at times that you have to crane your neck to hear, but something in her voice makes you listen. Pearson's life's work has been teaching dieticians to talk in such a way that those who are trying to lose weight – or stick to a gluten-free diet or manage diabetes, or make any of those other daunting dietary adjustments that people find so hard – can succeed in changing their behaviour. Since the late 1990s, Pearson has personally given thousands of training courses for British dieticians. Her work has taught her that whether someone changes their diet or not can come down to something as apparently trivial as how they are spoken to by members of the medical profession.

It is a bright early summer day. A circle of around fifteen women are sitting on chairs in a community hall, drinking coffee. 'I had to confess I found it hard to change the habits of a lifetime,' one woman confides. She is not talking about overeating or sitting on the sofa all day watching soap operas. This is phase two of one of Dympna Pearson's courses for healthcare professionals. The woman speaking is herself a dietician and is alluding to the habits of doling out weight-loss advice, without really listening to the client. 'Old habits die hard, don't they?' replies Pearson, with a look of respectful sympathy.

Dympna Pearson's firm belief, gathered from years of clinical experience, is that most dietary advice – however well meaning – is not just useless but counterproductive. 'One of our biggest banana skins,' she remarks, 'is persuasion.' Those who enter the dietetic profession tend – for very good reasons

– to have a fierce desire to change other people. It's frustrating to be sitting in a room with a morbidly obese person who just cannot manage to lose weight, or even maintain much motivation to do so, even though if they carry on eating as they do, they are heading for gastric band surgery. There is an itch to fix the situation. There's a tendency to come out with what Pearson calls 'all these lovely persuasive phrases'. 'Why don't you use a smaller plate?' 'Have you thought of eating an apple instead of a bar of chocolate?' 'Maybe it would help if you chewed more slowly?'

It's not that these suggestions are bad ideas in themselves, but to offer someone advice on their diet in this way is to treat them like a wayward child and yourself like an adult with all the answers. The trouble with this kind of talk – of which friends and family can be guilty as much as doctors and dieticians and governments – is that however nicely it is dressed up, with smiles and fake modesty, you are telling the other person what to do. And human beings don't respond well to being bossed around, particularly when it comes to something as personal as what you put in your mouth. At best, such advice will render a patient passive, going along with what is being suggested without ever really taking charge of the situation for themselves. At worst, it will make them still more resistant to change than they were before, because when people are told what to do, they often do the exact opposite. Pearson has noticed that the client's response to a string of advice tends to be a series of 'Yes, but . . .' answers. 'Yes, but I can't afford to buy smaller plates.' 'Yes, but the work canteen doesn't sell apples.' 'Yes, but I'm busy and I don't have time to chew more slowly.'

Pearson trained as a dietician in Dublin 'many moons ago' when the traditional model was all about reading out a diet sheet and waiting for patients to follow it. Or not. If they didn't follow the advice given, it must be their fault. After she qualified, Pearson began working with diabetic patients and felt a zeal of enthusiasm to make these people's lives better, to help them follow a diet that would save them from the worst consequences of untreated diabetes, such as blindness or coma. Yet she found the conversations she had with her patients tended to be unproductive: she 'cringes' now to look back on them. There was often an uncomfortable air in the room that she couldn't quite put her finger on as she read out the diet sheet of forbidden foods. And, more often than not, her clients couldn't manage to stick to their diets. She would find herself getting increasingly annoyed with people who dug in their heels and simply refused to change – such as the diabetics who left the session, went home and gorged on sugary snacks. It was only when she went on a counselling course that she saw what had been missing: with all that advice-giving, there wasn't much time left to listen to what the patients actually needed or wanted. After the course, she rethought her whole approach to helping people to change their diets. From now on, it wouldn't be about advice or persuasion, but finding a way of talking that could help people change their own behaviour.

'Motivation' is a word that Dympna Pearson uses a lot, but her idea of motivational talk is the opposite of what the word usually implies. It is not about running on stage in a cloud of dry ice and bombast and hectoring people into submission. Most of what Pearson does – and teaches others to do – involves either remaining silent or quietly paraphrasing back to people what they have just said: 'reflective listening'. She is a big fan of a 1991 book called *Motivational Interviewing* by William Miller and Stephen Rollnick, though she had already worked out much of her basic approach when she first read it.[18] Miller and Rollnick developed 'MI' in the 1980s as a way of helping people with problem drinking. Miller, who was treating and researching alcoholism at the University of New Mexico, decided to collect data on which patients were recovering the best. He was surprised to find that he could predict two-thirds of the variation in how well alcoholics had recovered six months after treatment 'based on how well their counselor had listened to them'. The clients who worked with the most 'empathic' therapist all succeeded in managing their addiction, whereas only a quarter of those working with the least 'empathic counselor' got better. Talk of empathy is not touchy-feely in this context. It can be the difference between kill or cure.

By the time someone reaches a dietician for weight loss, they are often in a state of defensiveness and despair. They may feel that they have 'tried everything' for years if not decades, and that nothing can or will make a difference. Some claim they do not actually want to change: they say they like their comfort foods too much, they are too busy to exercise, and anyway, they only came along because the doctor told them they had to. In this state, advice is the last thing that will help. Although counterintuitive, Pearson's approach involves 'rolling with the punches'. Instead of arguing and telling someone that they are wrong to eat such a bad diet – which will only create more hostility – you say something like, 'Oh, so it seems as if it's hard for you to eat healthily right now,' or, 'Would you say you are pushed for time to exercise?' It doesn't matter if there are pauses, because that shows that the dietician is giving the other person time to reflect.

As the conversation continues, maybe you ask them how important they feel it is to change. This is the point where

Pearson gets very excited. If she hears the merest inkling of an intention to change, she echoes it back at the client. It might be nothing more than saying something like, 'Maybe I do need to lose weight,' or, 'I would quite like to get my diabetes under control,' or, 'I wish my children ate better.' To Pearson such statements are golden, because they signal an intention, however slight, to try something different. 'We let this change talk fly away, we miss it all the time,' she says. But if a counsellor can only hear it and repeat it back, the patient – just maybe – will see that they, and not the health professional, are the person who is calling out for change. 'It brings a smile to my face when I hear them soften and say, "Well, I suppose I could . . ."'

We are all ambivalent about change. Faced with a particularly tempting platter of freshly baked biscuits, we may feel a bit like Zerlina in Mozart's opera *Don Giovanni*, who struggles – unsuccessfully – not to be seduced while singing the line '*Vorrei e non vorrei*': I want to and I don't want to. A person can passionately want to lose weight but equally passionately desire the comfort of a soft pillowy hamburger with all the extras. It would be dishonest to pretend that there are no downsides to eating slightly less than you could every single day. But the part of us that wants *not* to eat the hamburger or the plate of biscuits is also real. When a dietician hears the first inklings of 'change talk', says Pearson, they should not hurry the person on to the practicalities of dieting or an exercise programme, but try to capture the desire for change in such a way that the patient can hear what they have said. The dietician's job is not to persuade but to strengthen someone's own desire for change. At first, a client might say, 'I want to, but I can't.' Or they might say, 'I know I should,' which still has a tentative air to it. If the dietician can wait patiently enough, the person may nudge themselves out of their own ambivalence. Dympna Pearson sees her task as enabling someone to move from saying 'I want to' or 'I should' to 'I will'. This, to her, is the most powerful phrase because it signals a firm intention rather than just a vague inclination.

Pearson knows that this approach can sound 'touchy-feely'. But in her own mind it is simply good, evidence-based medicine. In fact, although there is as yet no conclusive evidence that motivational interviewing is the best way to enable dietary change, the indicators are encouraging. Four controlled randomized studies found that sessions of MI made people more likely to stick with a diet programme – whatever the programme might be – than conventional dieting interventions alone, involving advice, information and cognitive training in how to change behaviour.[19] There are signs that motivational interviewing can help people to maintain new eating behaviours long enough for them to become habitual. In one study, 148 obese women were all given a year of intensive diet treatment.[20] Each of them participated in eighteen group sessions aimed at giving them the skills and information to make major dietary change. Half of them also received just three sessions of individual MI with a dietician. A year on, the women randomly assigned to the MI group had lost 2.6 per cent more body fat than the others.

A small 2014 trial also found that motivational interviewing helped obese and overweight children lower their BMI.[21] Trials involving addiction rather than diet, however, have been mixed, with motivational interviewing proving more effective in some clinics than in others. Miller and Rollnick put this down to 'differences in clinician skill in delivering MI'.[22]

What is starkly clear is just how ineffective the old advice-giving is. As one article on motivational interviewing says: 'Confronting clients can lead to defensiveness, rapport breakage and, ultimately, poor outcomes.' When you hear Dympna Pearson modelling what the standard advice-giving conversation sounds like, it becomes obvious that – however good the intentions – this kind of talk will do no good at all. 'The essence of it is how you are with people,' she says. Several times, I saw Pearson do role-playing exercises to demonstrate how easy it is to slip into an unproductive conversation about change. To show other counsellors what is not effective, she keeps jumping in with 'helpful' pointers and talks more than she listens, with a very slight sharpness in her voice. Even though it is only role-play, you can hear the other person getting defensive and annoyed. It is painful to watch. It reminded me of the futile conversations I frequently have with my teenager, the ones where I confront him about not leaving his socks on the floor or clearing out his lunchbox; in these exchanges nothing is achieved, except that both of us are left in a slightly worse mood than before. The more someone is pushed, the more they come up with reasons why they can't or won't change.

It is a truism that no one can make someone else change. 'We shouldn't shove people into the swimming pool,' says Pearson, 'when they're not ready to get in.' Changing our diets always involves losses as well as gains. The water is cold when you first dip a toe in the pool. Giving up junk food involves a separation from some of your fondest childhood memories. Learning to like new foods can feel like leaving your old self behind. Recovering from an eating disorder entails giving up long-established coping mechanisms. Making yourself try foods that you find disgusting is, well, disgusting.

The best anyone can do from the outside is help a person get past their own ambivalence. If Dympna Pearson is right, the hardest part, after so many false starts and wrong turns, after all the failed diets and half-watched exercise videos, after all the stigma and the shame, after all the times you've told yourself that this diet would be different and all the times that it wasn't, is finding the motivation to get back in the pool and stay there long enough to get acclimatized.

Most public health campaigns aimed at changing diets are based on the idea that, once we are made to see that certain foods and behaviours are unhealthy, we will give them up. The evidence, however, shows that change in diet does not work like this. Whether you are a dietician sitting in a room with a diabetic patient or a government tackling an 'obesity crisis', persuasion does not cut it, because this is not how we learn to eat. At a social level, the key to improving diet is not pushing people to do something they are resistant to doing, but removing the barriers to change. These barriers could be psychological, cultural, economic or to do with the environment we live in. Sometimes our whole food system looks like a giant barrier to change, teaching us every day that it is normal to eat vast amounts of sugar and filling our heads with advertising images of beautiful healthy people eating unhealthy food. We speak of helping people to make better food choices, but in many modern food shops, to choose healthily involves ignoring nine-tenths of what is on display.

It looks as if the most common way individuals change diets across a lifespan is 'seamless change', which happens without conscious effort.[23] Examples of seamless change would be automatically buying more of something when the price comes down or unwittingly consuming different ingredients when manufacturers reformulate their products. From 2003 to 2010, the average salt intake in Britain fell by 15 per cent, not through individual choice but because food companies cut the amount of sodium in their products under pressure from lobbying groups and the government: a very benign form of seamless change.[24] The trouble is that most seamless change makes us eat less healthily rather than more so. Croissants start to slip into your mouth every day, as if by accident, because you get a new job where they are laid out with the coffee. Or you don't notice that the glass of white wine you always order is now much bigger and considerably more alcoholic than its equivalent a decade ago. A 2008 study of more than 400 people in the UK found that around 40 per cent were eating more takeaways and ready meals than they did during childhood, but most could not say why: 'The change just happened.'[25] By contrast, when someone tries to make conscious changes to eat more healthily, their path may be lined with hindrances.

Let's say you resolve to eat more fresh vegetables and fruit every day. You may never get beyond the planning and expense involved in shopping for them. One study found that resolutions to eat more bananas often collapsed at the first hurdle because there were no bananas in the house.[26] Even assuming you manage to lay your hands on a supply of fresh produce, there's the question of how to cook it. Among a sample of low-income families in Chicago, the ones who ate the fewest home-cooked meals were those who lacked the most basic kitchen equipment such as chopping boards, peelers and whisks.[27] And whether you have these items or not, you may not have much idea how to use them. Your plans to eat a wider range of vegetables may also be confounded by other family members who complain that they do not like them, in which case, do you cook yourself a separate meal or do you cook them for everyone and risk throwing good food away?

Culture is another barrier. As we've seen, traditional wisdom about feeding frequently clashes with the realities of our new food supply. In Britain, South Asians – from India, Bangladesh, and Pakistan, for example – make up the largest ethnic minority, and they are also statistically the ones with the greatest risk of heart disease and diabetes. Research has shown that among British Asians there are multiple obstacles to engaging in healthier behaviour.[28] Among the older generation especially, there may be a fatalistic attitude to illness: the view is that diabetes has been caused by fate, or Allah, or the dreadful British climate, so nothing can be done. Exercise in a gym is viewed by some Muslim Asians as individualistic and selfish and may be problematic for women in particular, given cultural expectations in many families that they should not sweat; or be seen to hurry; or wear sports clothes. As for food, the notion of eating smaller portions and limiting rich foods is antithetical to beliefs about hospitality. 'Indian sweets are supposedly for special occasions,' says Baldeesh Rai, a dietician who works with South Asian communities, 'but in an Asian household, anything can be a special occasion.'[29] Rai has found that in many South Asian families, it is possible to change diets only if the cook of the family – often the mother-in-law – is involved. It doesn't matter how much information you have on the calorie content of ghee if you are not the one measuring it out.

If you spend too long thinking about all the barriers to change it would be easy to endorse the common view that almost no one really succeeds in long-term weight loss. You may start to feel hopeless about your chances of sticking to a diet, if that is what you are trying to do. The general opinion is that you might be able to shed a stone or so in the short term, but you will put it all back on again, with interest, and be worse off than you were before. Anyone with weight issues – or so the thinking goes – is fated to grapple with them for a whole lifetime, without much possibility of improvement. This is a very depressing thought, particularly if you are unlucky enough to have become obese as a child.

Fortunately, it isn't true. No one could pretend that losing weight and keeping it off is easy, but the evidence suggests that around 20 per cent of overweight dieters – one in five – do in fact manage long-term deliberate weight loss, defined as losing at least 10 per cent of initial body weight and keeping it off for at least a year.[30] Relatively few studies on weight loss have followed participants for long periods of time, but those that have indicate a sizeable minority do succeed in losing weight without regaining it one year on; and three years on; and even five years on. A little-publicized piece of good news is that, over the past two decades, long-term maintenance of weight loss has improved for those most in need of it. Dr James Anderson is an endocrinologist at the University of Kentucky. Anderson has found that, compared with the 1990s, more of his severely obese patients are able to maintain a large weight loss, perhaps thanks to the use of more intensive and frequent sessions of behavioural coaching.[31] It has proved possible for some patients who needed to lose more than 100lb – and who might otherwise have been candidates for bariatric surgery – to achieve sustained weight loss (with follow-up after five years) using meal replacement shakes, carefully managed main courses and lots of fruit and vegetables, plus regular medical support.[32]

The vital question is what is it about the successful 20 per cent – referred to as weight 'maintainers' in the literature – that makes them able to lose weight and keep it off? They seem to have certain habits in common that set them apart from the relapsers. One factor is that maintainers are much more likely to engage in regular exercise, ideally an hour or more of moderate physical activity every day. This pattern has been confirmed by numerous studies: relapsers do not exercise consistently, whereas maintainers do.[33] We do not know whether exercise helps prevent relapse because of the expenditure of energy, because it's time when you are not eating, or because it contributes to wellbeing: the dopamine and serotonin released when exercising can help prevent depression. Of course, it may also be that the kind of people who persist with weight loss may also be people who persist with exercise. Correlation is not necessarily causation.

There are certain other habits that maintainers have in common. A study of more than 4,000 of them found that they tend to eat breakfast every day and stick to a consistently moderate diet across the week and across the year, rather than holding back during the week and splurging on weekends and holidays (5:2 dieters – take note!). Long after the initial weight-loss 'diet' is over they continue to monitor what they eat and are flexible enough to deal with small slips before they turn into major relapses, without beating themselves up. Some of their success may be because of their emotional state. They are less likely to be depressed and they are much less likely to be binge eaters. 'Disinhibition' around food and emotional eating are strong predictors of weight regain. It is, as always with eating, hard to unravel the complex web of cause and effect here. Relapsers tend to have lower self-esteem and poorer body image than maintainers, but this may be precisely because when they look in the mirror, they feel bad about the weight they have regained. They also seem to feel worse about the whole process of eating.

Counterintuitive as it might sound, the maintainers enjoy their food more. This key difference between maintainers and relapsers was identified by a 1990 study from California.[34] The lead researcher, Susan Kayman, a public health nutritionist, noted that 'surprisingly little is known about those who lose weight and regain it'. Kayman decided to find out more by conducting in-depth interviews with three groups of mostly middle-aged women: formerly obese women who had successfully maintained weight loss; formerly obese women who had lost weight and regained it; and average-weight women who had neither put on weight nor lost it. The interviews revealed that in many ways the maintainers were not so different from the relapsers. They did not differ much in marital status or whether they had children, though they were slightly more likely to have a university education and to work outside the home. The most substantial difference was the way they ate. Maintainers told interviewers that they had never completely restricted their favourite foods and that they 'made efforts to avoid feelings of deprivation while changing food patterns'. As time went on, their appetites had changed. They no longer wanted to eat such large quantities and many had lost their taste for sweets and doughnuts, finding them excessively sweet and fatty. They had changed the way they cooked, not frying food as much as before or using as much sugar, and they included more fruits and vegetables and made smaller portions. But the real change was to themselves, because this was how they now wanted to eat. Like the Japanese, they did not start off eating well; but they were able to change their habits and preferences until they got to a point where delicious food and healthy food were one and the same.

By contrast, the relapsers associated weight loss with eating foods they did not like. While the maintainers devised diet plans to suit their own lives and tastes, the relapsers tended to follow rigid diet programmes that actively went against their own food preferences. While 'on a diet', the relapsers would forbid themselves from eating anything that they actually enjoyed. As Kayman puts it, they 'perceived their diet foods as special foods, different from the foods their family could have and different from the foods they really wanted'. The whole time they were eating these foods, they felt deprived. It didn't take much for them to give up the struggle and return to the old patterns of eating. Seventy-seven per cent of relapsers interviewed by Kayman said that the trigger for weight gain was life throwing up some kind of complication that made them revert to their normal foods.[35] The major obstacle to dietary change is in some ways the most obvious one: no one – adult or child – wants to eat foods that they do not like.

Though this sounds obvious, it stands in contradiction to the way that almost all our schemes for healthy eating up till now have been constructed – whether on a personal level or a social one. Adam Drewnowski, a professor of public nutrition who studies ways in which diets can be improved in whole populations, notes that 'nutrition education and intervention strategies aimed at improving diet quality have focused almost exclusively on the nutritional quality of foods and not on the taste or pleasure response'.[36] This is a huge wasted opportunity because nutrition will only improve if you get people to consume healthier food. And people will only consume healthier food over a lifetime if they consistently choose to eat it. Other things being equal and assuming that healthier food is available and not too expensive, you will only choose it if it's something that you enjoy. Instead of intervening at the level of nutrition and information – eat less sugar! – a better place to start would be pleasure. This can be illustrated with a chart. The following shows, roughly speaking, how we arrive at the nutritional advantages of eating healthy food. Let's take broccoli as an example.

1. SENSE. You see, smell and taste the broccoli: its green colour, its sweet crunchy stalks and soft fluffy florets.

2. RESPONSE. You respond to the broccoli. Maybe with pleasure, maybe with pain. Your response here will be influenced by whether the broccoli is forced upon you or offered with enthusiasm and the skill with which it is cooked, as well as whether you are a 'bitter taster' and how many times you have been offered it before.

3. PREFERENCE. Based on your response, you form a preference. Either you become a broccoli lover or a broccoli hater, or somewhere in between.

4. EAT. This preference decides whether you regularly choose and eat broccoli; or not.

5. NUTRITION. Whether you eat broccoli will decide whether you gain all the nutritional advantages of eating it, which include folate, fibre, vitamin C and calcium along with certain disease-fighting phytochemicals.

There is small chance of getting the health benefits of broccoli unless we take the right path through 1, 2, 3 and 4.

It doesn't matter how much 'nutrition' there is in a given food unless someone puts it in their mouth. Public health campaigns and diets alike almost always start at stage 4 or 5. We are told how many benefits there are to eating green leafy vegetables and urged to eat more of them. When we fail to change our behaviour, they tell us again. And then again. But no one thinks to check first whether we like green leafy vegetables or whether we have even tasted them. In 2010 Jamie Oliver's TV series *Food Revolution* revealed that many children could not correctly identify many raw vegetables by sight, including potatoes, cauliflower, tomatoes, beetroot and aubergine. This suggests that the adults in their lives had never learned to like and cook these vegetables either. You are unlikely to eat something if you don't know what it is. It's like being pushed into the swimming pool. The real aim should be to get people to like healthy eating sufficiently that they jump in of their own accord. By the time we are at stages 4 and 5, it's too late. For a real change in our diets to happen, we need to go back to 1, 2 and 3. When our preferences are in order, nutrition should take care of itself.

Earlier on, I asked what it would take for us to enjoy a 'hedonic shift' to enjoying real, whole, food. It will perhaps come as no surprise by now that the answer is frequent, positive exposures to those health-giving foods. The surprising part is what a short time-frame may be required to shift our palates in a healthier direction. Our tastes are built over decades and reinforced daily by meals and snacks. Yet experiments have shown that at least some of our flavour responses can be relearned over a matter of weeks. The olfactory system is one of the very few parts of the adult brain that is constantly regenerating itself. The brain is flexible enough (the technical term is 'plastic') to change its responses to flavours

over a very short period of exposure. This has been demonstrated with salt and sugar, which you might expect to be the
most immovable of our tastes.

If we consistently eat less sugar, it actually changes our
sense of sweetness. In the late 1990s biologists at Clark
University in Massachusetts started experimenting to find
whether being intensively exposed to fructose or glucose could
affect an individual's ability to perceive low concentrations
of other sugars. They found that just five short exposures
to glucose over a few weeks could make individuals more
responsive to sweetness in very weak sugar solutions.[37] The
good news, however, is that the effects were reversible. After
the experiment was over, the subjects returned to their normal responses to sugar after just a few weeks. This suggests
that, if we could only take a fortnight's holiday from sugar,
we might return to it less fondly.

The same is true of salt. Experiments suggest that reducing
salt in the diet for a period of just eight to twelve weeks is
enough to reduce the pleasure of eating very salty foods.[38]
Interestingly, it seems to take hypertensive (salt-sensitive)
people longer than others to kick the salt habit, though it
is not clear why this is so. But a study of both normal and
salt-sensitive adults found that after three months on a low-
sodium diet, 'a significant hedonic shift occurred' in all the
adults. Before the experiment started, they all rated salty
foods as more enjoyable than non-salty ones. After twelve
weeks, this changed. The subjects no longer found reduced-
sodium versions of chicken broth, crisps and crackers any
less pleasurable than the 'normal' high-sodium varieties.[39]

By making enough of these tweaks to our diets, we may
reach the happy state where the foods we crave the most –
give or take the odd French fry – are ones that do us good. It

is possible to go back to learning to like basic healthy foods just as we did when we were children. As Dr Spock wisely remarked in his bestselling *Baby and Child Care* from 1946: 'Feeding is learning.'

My daughter has a friend called Lily. She used to be one of the fussier children we knew. She couldn't stand 'mixed-up' food or anything that came in a sauce. The main things she liked were meat and potatoes and plates of plain cut-up cucumber. Not only could she not stand to eat a tomato, she couldn't bear to have so much as a trace of a tomato on her plate. This prevented her from trying most pasta dishes; and salads; and curries; and stews; and her mother's home-made pizza. She also wouldn't eat any fruit, except for raspberries. This was difficult both for Lily and for the rest of her family, who are adventurous eaters and lovers of spicy Indian dishes such as sag aloo made with spinach, potatoes, ginger and tomatoes. Often, she ended up eating separate meals of fish fingers and chips. There seemed to be no way out of her limited diet.

Then, aged ten, she was casting around for a good New Year's resolution and suddenly decided to do something about her restricted tastes. The idea was her own; her parents did not pressure her into it. Lily – a sunny, chatty person – set herself the task of trying one new food each month. At the end of the month, she still might not like the new food but at least she would have given it a go. Somehow, the spirit of fun and adventure in this little project made it possible for her to put foods in her mouth that previously she would have recoiled from. It was the opposite of most grown-up New Year's resolutions, which tend to involve cutting things out rather than adding them. Whenever we saw Lily that year, she was excitedly talking about the food of the month.

The very first month, she successfully taught herself to like home-made pizza, despite the fact that it combined cheese and tomatoes, two of her no-go foods before. In subsequent months, she learned to eat chicken curry and apples and spaghetti bolognese and meat in sauces. By the end of the year, she still wasn't crazy about bananas or salads or any kind of fish except for fish and chips. But in just twelve months she had vastly expanded her repertoire of foods and had also proved to herself that it was possible to enlarge her world of eating, whenever she needed to.

Because we live in Britain, Lily's year-long resolution was seen by her friends as rather unusual; maybe even a bit weird. In Finland, however, this kind of sensory exploration has now become a basic part of every child's education. Lessons on taste are also given in schools in Sweden, Denmark, the Netherlands and some parts of Switzerland and France. This education in eating is part of the growing 'Sapere' movement. In Latin, sapere means both 'to taste', 'to be able' and 'to know'. The idea behind Sapere is that it is possible to educate children in the pleasures of food; and that doing so will set the children up for a lifetime of healthy eating. Feeding is learning.

The inspiration behind teaching children about taste – which may not amaze you – is French. In France, more than elsewhere, there has long been a deeply cherished belief that a child's education involves being 'civilized' in the pleasures of the table. In a famous experiment in the nineteenth century, a doctor called Itard brought a wild boy into his care. He named him Victor. For twelve years, Victor had been living in the woods of Aveyron and at first he only wanted to eat the fruits of the forest to which he was accustomed. But over time, Itard succeeded in 'awakening' new tastes in the boy for 'a whole quantity of dishes he had hitherto always disdained'.[40] Dr Itard indoctrinated Victor in the joys of French cuisine, which he saw as a passport to civilization.

More than a hundred years later, another French scientist called Jacques Puisais had a similar idea. Puisais (born in 1927) is a chemist and wine obsessive who believed that children should and could be trained to become more discerning gourmets. Puisais worried that new generations were growing up without the ability to respond to complex flavours or to appreciate the finer points of cuisine. He founded the French Institute of Taste and in 1974 he started the first 'taste education classes' in French primary schools: *les classes du goût*.[41] A typical programme in French schools based on Puisais's ideas would start with the five senses, move on to knowledge of French regional specialities and culminate in a grand 'festive' meal at a fancy restaurant where pupils would learn table manners and the art of *savoir-vivre*.[42]

You can see why this very French vision of food education did not immediately catch on elsewhere. Outside France, to say that someone is not 'educated' because they do not know how to sit at a table and eat a three-course restaurant meal with the right cutlery might sound ever so slightly . . . how shall I put this? . . . snobbish. But nutritionists and educators elsewhere in Europe took Puisais's original vision and developed it in ways that were more democratic and more directly aimed at improving health. From the 1990s onwards, schools in Sweden started offering classes on 'food for the senses' and the Netherlands followed in 2006. But the country that has most fully embraced Sapere is Finland.[43] From 2009 to 2014, the Finnish government took the ambitious step of funding Sapere food education in all kindergartens in the country. There are now more than 7,000 professionals in Finland trained in Sapere methods. This is by far the largest experiment ever conducted into changing children's tastes for the better.

The Finnish engagement with Sapere was prompted by a national 'alarm' in the early 2000s about children's eating habits. Finland had notably higher child obesity levels than the country's neighbours in Norway and Sweden (9.2 per cent of Finnish boys were obese as against 5.1 per cent in Norway and 4.2 per cent in Sweden). Nursery school teachers noticed that many of the children in their care were consuming large amounts of sweet foods and drinks, and not many fruit and vegetables: this now familiar story. They feared the children were facing a future of health problems. Nursery staff also noticed that the children's eating habits were heavily influenced by their family background. If a change was going to come, it would need to be from the school rather than the home.

Sapere was first tested in Finland in the rapidly growing city of Jyväskylä, a lakeside city where the winters are long and cold and it's tempting to stay inside eating sweet cardamom buns filled with jam and whipped cream.[44] In 2004–5 the kindergartens of Jyväskylä received funding to give all children in the city aged one to seven lessons in nutrition and 'varied food habits'. The goal was to create a 'positive and natural relationship with food and eating'. A team of nutrition researchers advised the kindergartens that to do this, they must abandon the dogmas their own parents had taught them, such as 'Eat everything on your plate' or 'Don't play with food'. Instead, children would be positively encouraged to play with their food by exploring ingredients with all their senses: the hard crackle of rye crispbread, the soft fuzz of a peach, the puckering sourness of raw cranberries. This sense of exploration continued in what they ate at lunchtime. 'Pedagogical' menus were served composed of foods that the children were learning about, with more emphasis on vegetables and fruit. Nursery workers found that small children would eat many more vegetables if they were allowed to pick them up with their fingers.[45]

The results from Jyväskylä were so promising that Sapere was extended across all Finnish preschools. Teachers reported that by including food in the children's education each day, their attitudes to eating could be radically altered. During the Sapere project, children 'dared to test stranger foods' than before. Parents were surprised to find that their children had acquired new knife skills, more varied tastes and fresh attitudes. Instead of being disgusted by beetroot, they were now fascinated by the question of how it turned the cooking water purple. The children were more conscious of what they ate, and more attuned to whether they were hungry or full. Most strikingly of all, there are hopeful signs that Sapere has led to a reduction in the incidence of child obesity in Jyväskylä.[46]

This happened not through explicit lectures on nutrients but by channelling the children's natural curiosity. The change was largely unconscious. Arja Lyytikäinen, a nutritionist who oversees the Sapere programme in Finland, says it is all about 'learning by senses, learning by playing'. Some days, the children might go berry-picking; other times, they make bread, chop fruit for fruit salad or draw pictures of vegetables. Most of the time, they are hardly aware they are learning at all. Sometimes, they play the 'lemon thief game' where one child leaves the room and another child rubs lemon on their hands. The 'detective' comes back in and has to say who has been stealing lemons from the garden. Many of the Sapere sessions take the form of open-ended sensory games, where the children describe the sight, taste and smell of different foods. They discuss whether they prefer to eat carrots raw or cooked; whether they enjoy bread more with garlic, with butter or plain. In one Sapere session in Jyväskylä, a child observed that white pepper 'attacks the nose'. Another said that blue cheese was 'soft, white and green . . . just like a ghost'.[47]

In earlier generations, to criticize food like this would have been seen as bad manners. But the prospect of saving a generation of children from bad diets and ill health trumps any niggles over etiquette. The aim of Sapere is to get children to know their own authentic tastes. 'Everyone has his own preferences', is one of the mottoes; another is 'Matters of taste are not argued about, but discussed'. Each child is encouraged to design their own birthday cake, including the fillings and decoration they love the best. Like Dympna Pearson, the Sapere teachers in Finland have found that dietary change comes not from forcing someone to eat what they do not like, but in helping them to discover their own passions. In a Sapere-run kindergarten, the children have many different tastes. Some prefer blueberries, while others like lingonberries. Some gravitate towards sour; others like saltiness. But, like the children in Clara Davis's experiment in Chapter One, every one of them ends up with a set of taste preferences varied enough to enable them to eat well when they are older. What Sapere shows is that any child can learn to eat better, given the right encouragement and access to a variety of good foods.

The changes to eating behaviour set in motion by this type of 'sensory education' are profound. It is not about learning to like this or that vegetable; but developing an overall attitude to eating that is more open to variety and less governed by the simple sugar-salt-fat palate of junk food. Like the successful weight-loss maintainers, Sapere children no longer respond to the simple sweetness of sweets and soft drinks in the same way. They start to hanker after the 'feistiness' of lemon and the earthy taste of rye crackers. A series of studies by the psychologists Hely Tuorila, E. P. Köster and others have shown that giving sensory education to children aged around eight to ten can leave them with much more positive responses both to novel flavours and to complex ones.[48] Köster has demonstrated that one of the effects of a sensory education is to give children a taste for more complex food. They start off preferring simple flavours, but after the sensory education, the complex flavours – what children call 'mixed-up' food – tend to be the ones that are loved best of all. They start to prefer mashed potatoes with celery and nutmeg to plain unseasoned potatoes.[49]

Best of all, a sensory education seems to have the potential to free up a child from many of the old barriers to trying new food. Tuorila – who advised the Finnish government on Sapere – has done experiments showing that sensory education can make children feel much more favourably towards all unfamiliar foods, not just the ones they tasted during the lesson. Tuorila notes that neophobia is usually considered to be a personal trait that will never alter. Around 40 per cent of Finnish adults say they do not like many vegetables because they have never tried them. And yet this attitude can and does change, even with those who have an individual disposition – like my friend Lily – that makes them inclined to reject new and complex flavours. According to Tuorila, a Sapere-style education shows that it is possible for children to learn to improve their eating skills in ways that will automatically lead them to a healthier diet.[50]

No person is destined by genes to eat badly (they may be

trapped by poverty or neglect, but that is another matter). Nevertheless, many of us do seem to be stuck when it comes to food. We are stuck in habits and attitudes that seem impossible to break. We are stuck thinking food is love. We are stuck with guilt about food because we are female; or stuck not liking vegetables because we are male. We are stuck feeding hungers that often exist more in our brain than our stomach. We are stuck in our happy childhood memories of unhealthy foods. But the biggest way we are stuck is in our belief that our eating habits are something we can do very little about. In fact, we can do plenty. The first step is seeing that eating is a skill that each of us learns and that we retain the capacity for learning it, no matter how old we are.

Sapere shows how changes to personal eating habits can happen in tandem with changes to the national food culture – such as those seen in Japan at the start of the chapter. In an ideal world, other countries would follow Finland's example and recognize that learning to eat a good and varied diet is a key part of every child's education. Arguably, the consequences of not learning these skills can hamper a child's future as badly as growing up illiterate and innumerate; worse, factoring in the damage to health. Early childhood, as we have seen, is the time when we are most receptive to developing new tastes. I ask Arja Lyytikäinen, the main nutritionist overseeing Sapere in Finland, what the optimal age is for offering this type of food education and she replies that they have found it works best with children aged one to six, or maybe as old as ten. The bitter and sour tastes are easier to acquire when you start young and the younger the child is, the more chance there is that their own education in food will help the whole family to improve the way they eat. There's also the fact that children are more open to learning.

But it's never too late. Arja Lyytikäinen says that in Finland they have had some very positive results using versions of Sapere with adults. Mental health clinics have used sensory food education during group counselling; it has also been used to help teenagers with type 1 diabetes to improve their diets. There have even been small experiments in Scandinavia with 'taste schools for the elderly', in which, just like the pre-school children, those nearer the end of life learn how to explore new foods. Beyond the Sapere system, interventions with the elderly in Canada have suggested that taste workshops are a more effective way to teach about nutrition than other methods such as leaflets or lectures, which can make older adults feel patronized.[51] As many as a third of those in care homes are malnourished; inadequate intake of protein, vitamin D and fresh vegetables are a particular problem.

In old age, without the distractions of work to fill the day, food becomes an ever more central preoccupation, yet the barriers to eating well only increase. Impairment to the sense of smell and taste can make meals seem insipid. Difficulties with swallowing are very common. Poor handgrip and frail arms may impede cooking. Yet the greatest obstacle remains the old familiar one, unchanged from childhood, of summoning the appetite to fulfil nutritional requirements. A 2004 study of older British men living on their own found that only 13 per cent managed to eat the recommended '5-a-day' of fruit and vegetables.[52] One participant, aged seventy-nine, remarked, 'I don't eat greens or fruit, I hate them, I hate greens.' It horrified this widower to step foot in a fruit shop, though his grandchildren wished he would try. He had eaten vegetables from a sense of obligation when his wife was alive but now that she had died, he bought no vegetables 'whatsoever'. Just like people with eating disorders or weight problems, the elderly often need to relearn their responses to the foods that will nourish them. Nurses and other health-care professionals working in geriatric care are inclined to dismiss the under-nutrition of the elderly as something they are powerless to help, precisely because it is so common. In 2006–7, however, a group of Swedish researchers decided to test whether it might be possible to increase the enjoyment of healthy food in old age through a 'taste school for the elderly'.[53] The initiative was carried out in Skane county in southern Sweden. A cookery teacher offered Sapere-style education to a group of twelve people with an average age of seventy-five: eight women and four men, all of whom lived alone. The idea was to 'increase their lust for cooking and enjoyment of healthy meals'. They were also taken on guided walks by a physical trainer.

Unlike other Swedish nutrition initiatives for the elderly, which have tended to emphasize health, this one started with 'joy'. Not that everyone in the group saw it this way at first. Three of the men said they had no interest in changing how they ate. With an average age of eighty, who can blame them? Yet after three months, all the participants said that the programme had taught them 'a lot' and increased the pleasure they felt in cooking for themselves. Each food session began – true to the philosophy of Sapere – with sensory training. On one occasion, the septuagenarians were asked to taste solutions containing sweet, sour, salty and bitter flavours before going on to prepare a bitter chicory salad with garlic dressing, and meatballs with berry sauce served with vegetables cooked with and without salt.

Through the cooking and taste sessions, this small group started to relish the perfume of spices that they had never imagined liking before. They discovered a fondness for vegetables such as fennel and sweet potato which they had gone seven decades without knowing. Twelve people is a tiny sample, and sadly, despite its positive outcome, the project was not repeated. The lead researcher – Kerstin Ulander – died the year after the experiment was completed and Sapere methods have not been widely disseminated within geriatric care in Sweden or elsewhere. A colleague of Kerstin Ulander's told me that it was 'surprising' that Sapere is not offered more widely among the elderly.[54] The reason, he suggested, was 'lack of knowledge' among clinicians working in geriatric care. Yet this project – however modest – offers yet another hint that, under the right conditions, food habits can be changed for the better, at any stage of our existence. Even towards the end.

In contrast to all the other things we work on in life that are far less likely to increase our wellbeing – including dieting – it is astonishing how little effort we put into changing our eating preferences for the better. There is every indication that the basic methods of eating better – increasing variety, including more plant foods, structuring meals and becoming more responsive to hunger cues – can be learned at any age, given the motivation. In the preceding chapters, we have seen that while we think of our 'tastes' as being an intimate part of ourselves, they are actually mostly learned; and therefore can be relearned. Remember how the great Karl Juncker – in Chapter One – taught himself in his thirties to like the strange substance called salad cream so that he would be able to consume British salads. Or think back to how Keith Williams, in Chapter Seven, was able to use his Plate A-Plate B system with both adults and kids to free them from a lifetime of picky eating.

You were a child once, too. When you arrived in the world,

your only food preferences were milk and buried memories from your mother's diet. Those early weeks were dominated by meals – the stab of hunger, the sweet contentment of being sated – but you could not yet tell dinner from breakfast. You didn't yet know – lucky you! – what a trans fat was; or a frappuccino. No one had taught you to worry whether you were getting enough protein, or to feel guilt when your stomach was full. You had never watched a fast-food commercial, and on the relative merits of quinoa and *macarons* you had no opinion. Food was wide open for you. The great garden of ingredients – from bitter greens to sweet dates – was all equally unknown: all new, all strange, all waiting to be discovered.

It may not feel like it, but you never lost your potential to change how you eat. The wonderful secret of being an omnivore is that we can adjust our desires, even late in the game. It won't happen on the first bite. Long-standing appetites do not take kindly to being ignored; it is unsettling at first to lengthen the gap between meals or to leave your habitual foods uneaten. Sometimes, it is hard to wrestle your disgust long enough to put something new in your mouth. Assuming that you don't retch or die, maybe you try the experiment again. Over time, you forget that this food was ever strange. It settles into something like a pleasure. One day, you eat a plate of cucumber with mint and instead of finding it dull, you marvel at how clean and herbal it tastes. Now, it is the old hungers and habits – the sickly sugar headrush, the lingering salty aftertastes – that feel uncomfortable. Given enough repetitions, the new ways of eating may become as familiar and sweet to you as milk.

CHILLI

If you still doubt that tastes are something we learn, consider the chilli pepper. On first encounter, chilli is irritating, containing a chemical (capsaicin) that activates pain receptors on the tongue. Chilli burns! Yet in many parts of Asia, Africa and South America, these pungent peppers (*Capsicum sp.*) are eaten with relish every day.

The obvious assumption, which those of us with milder diets leap to, is that chilli just doesn't feel as painful to those who eat a lot of it. But a groundbreaking paper by Paul Rozin and Deborah Schiller in 1980 refuted this.[55] Rozin and Schiller tested Mexicans in a rural village where chilli is eaten three times a day. Surprisingly, they found that the Mexicans were not desensitized to chilli. They felt the burn just as strongly as Americans who ate chilli only once a week on average. The difference was they enjoyed it more.

Very young children are generally protected from the burn in chilli-eating cultures. The only time a toddler is likely to taste chilli is when a mother places chilli on her breast to wean the child off her milk, a gesture that confirms that children find the spiciness horrifying. The question, as Rozin and Schiller put it, is 'How do tens of millions of little chilli haters become chilli lovers every year?' In the Mexican village they studied, almost everyone over the age of five or six ate chilli in some form at every meal. The villagers yearned for it when it was not available and said that food tasted bland without it.

Rozin and Schiller argue that the human conversion to chilli – which is not seen in other omnivores such as rats – represents a 'hedonic shift'. Around the age of five, children start to season their own food. They see older siblings and parents reaching for the salsa on the table and start to copy them. Maybe the first bite makes them cry with pain, but over time they start to enjoy the after-effects. They associate chilli with other good-tasting foods, such as tortillas and refried beans. They develop a perverse enjoyment of the very aspect of the chilli that they first disliked: the warmth and pain it causes in the mouth. Rozin compares it to the benign masochism of watching horror movies or going on rollercoasters.

Not everyone needs to learn to like chilli. Some people never stop finding it irritating. But for chilli, we could substitute many other pungent foods, from bitter vegetables to sour citrus to strong cheese to peppery olive oil. The fact that millions of children every year can learn to like chilli offers hope for us all; that our next bite can be different from the first.

EPILOGUE:
This is Not Advice

After everything I've written about the futility and inefficacy of so much dietary advice, it would be pretty rich for me to give you any. No amount of urging from me to eat this or that food will make you eat it. If I suggest you give up the foods you love, you might – quite understandably – tell me where to get off. So I won't try. I have no idea what your personal circumstances are. I don't know what's in your fridge or what your views are on cheese or whether gluten agrees with you or whether you run marathons or how easy you find it to decline a second slice of pie or whether your mother gave you sweets when you cried. Maybe you are one of those fortunate people for whom neither food nor weight has ever been an issue. Good luck to you!

But in the course of writing this book – not to mention through the experiences of eating food for forty years and feeding various people, including my children – I feel I've learned some things that have helped me to make my peace with eating. These are some of the insights I wish I'd figured out sooner. I hope you won't mind if I pass them on.

- Eating well is a skill. We learn it. Or not. It's something we can work on at any age.

- Sugar is not love. But it can feel like it.

- No one is doomed by genes to eat badly. Pickiness is governed more by environment than biology.

- We mostly eat what we like (give or take). Before you can change what you eat you need to change what you like. And you will never like new foods unless you give yourself the chance to try them. The fact that you don't like something now is not necessarily a sign that you will never like it.

- If you want yourself to eat better, focus less on the food and more on your own response to it.

- Nothing tastes good when it's eaten in a spirit of coercion. The secret is – as far as possible – to make healthy food and pleasurable food one and the same.

- Your first job when eating is to nourish yourself.

- Most people eat much better when most of what they eat is home-cooked. But this principle only works if you learn to make at least a few things other than cupcakes.

- No one is too busy to cook.

- Disgust is even more powerful than desire. We should use this more to our advantage. Become a food snob. The ideal scenario for healthy food shopping is when you won't buy most of what's for sale, not because you shouldn't, but because it repels you.

- Calories are not the same as morals. No food can be either 'naughty' or 'virtuous'. It's all just food.

- Before you change what you eat, change how you eat. It is virtually impossible to have a healthy relationship with food if you don't eat structured meals. I won't tell you how many meals a day you need. It might be two large ones; it might be five or six smaller ones. Either way, don't skip them.

- Eat soup.

- If it's not a mealtime and you are wondering which of two 'healthy' snacks you should buy, the answer is probably neither.

- If it is a mealtime, and you are vacillating between two main courses, go for the one you really love. And when you are full, stop.

- No one likes waste, but it's time to abandon the idea that it's bad manners to leave food on your plate. What's bad manners is to make someone feel ashamed for leaving food on their plate when they are full.

- Smaller plates – and smaller lunchboxes and smaller wine glasses – really do work (and mean you can minimize wasting food when at home). Eat dinner on side plates or bowls and dessert on saucers. If you spend a week or two weighing everything you eat on digital scales – without actually calorie counting – you start to see how out of kilter our idea of portions has become.

- Rethink what counts as a main course. Instead of having a large pizza with a tiny salad garnish, have a huge salad with a small pizza on the side. It's still a very comforting meal.

- Not every happy occasion needs to be marked with a gargantuan cake smothered in sugar. You may find equal joy in a smaller cake, a punnet of cherries and a victory dance.

- When making small changes to your diet, try to avoid the mindset of deprivation. Except for my first coffee of the day, I recently switched from milky lattes to black coffee. Instead of feeling sad that I wasn't drinking all that foamy white milk, I would ask myself whether I'd rather have a drink of water or a drink of black coffee. I chose the black coffee. It then tasted so much nicer. Clearly, the next thing I need to work on is my caffeine addiction.

- We hear a lot about 'superfoods'. The term is used for foods that are supercharged with certain nutrients. Usually, this is a marketing device, trying to get you to hand over your cash for expensive exotica, such as goji berries or wheatgrass. But how many people do you know who actually eat goji berries on a daily basis? The real superfood would be one that you enjoy that also happens to be healthy: crisp, sweet apples, say; or hard-boiled eggs with celery salt for dipping; or warm asparagus with sesame-soy dressing; or Moroccan carrot and garlic salad. The more of these superfoods you can build into your personal repertoire, the better you will eat.

- Regular exercise definitely helps: the endorphins, the expenditure of energy, the fact you are doing something other than eating. But again, find a version that you like so much you positively want to do it, rather than the one that burns the most fat but leaves you so drained you need to take urgent solace in carbohydrates.

- If you want your children to eat better, don't tell them what to do; eat better yourself.

- Most of our approaches to feeding children are too short term. We worry about the next five minutes when we should be thinking about the next five years. If you pressurize a child to eat a whole plate of greens, you are teaching them to dislike the greens – and you, for that matter. If you persuade them to take one tiny taste (and again tomorrow and the next day and the day after that), there's a chance they will become a lifelong eater of greens.

- Cajoling, urging and hinting doesn't change how people eat. It doesn't work with children and it doesn't work with adults.

- Girls eat better when food stops being something forbidden.

- Boys eat better when their parents continue to expect them to eat vegetables and include them in home-cooked meals as they get older. Or – better still – get them to prepare the home-cooked meals themselves.

- It is genuinely possible to reach the point where you desire broccoli more than fries and wholemeal sourdough more than sliced white bread.

- Hunger is not always a signal to panic. A day in which you haven't had a couple of spells of feeling slightly peckish is probably a day – sad to say – when you ate too much.

- No one is truly omnivorous. You are allowed to find some foods revolting. You don't have to like sprouts. It's when you don't like any or many vegetables that you are in trouble.

- Changing the way you eat is hard, but it can be done. Look at Japan.

Further Reading

The bibliography lists the sources I drew on when researching this subject. I'd like to mention in particular the multi-
faceted work of Paul Rozin, whose research interests in food
straddle psychology, culture and neuroscience and who seems
incapable of writing a boring sentence. If you're interested in
more practical day-to-day thoughts on eating better, I recommend the following books. What they have in common is
that, instead of doling out a set of rules on what foods we
should eat, they look in a more holistic way at methods and
approaches by which we can start to eat better.

Mindless Eating: Why We Eat More Than We Think by
Brian Wansink shows how much we delude ourselves when
it comes to how much we eat and provides useful techniques
anyone could use to avoid overeating. *VB6: Eat Vegan
Before 6:00 to Lose Weight and Restore Your Health* by
Mark Bittman describes the regime Bittman – the food writer
for the *New York Times* – adopted after a doctor warned
him he was pre-diabetic. Bittman now eats nothing but vegan
food until 6 p.m. and anything he likes thereafter. Even if
you don't wish to follow him down the full vegan route –
for me, breakfast toast without butter is too gloomy – his
'flexitarian' approach offers a pragmatic model of how you

can change your eating permanently without going 'on a diet'. *A Change of Appetite* by Diana Henry is a wonderful collection of 'accidentally healthy' recipes, none of which tastes like deprivation, interspersed with essays on nutrition; another cookbook that has helped me to eat inadvertently better is *A Modern Way to Cook* by Anna Jones, a collection of sumptuous yet light vegetarian recipes. Finally, *Child of Mine: Feeding with Love and Good Sense* by Ellyn Satter is full of wisdom about how to set children up with healthy eating habits, without mealtimes becoming a battleground. Satter writes about the aim of feeding as being to enable children to master certain 'competencies'. These include: to like eating and enjoy being at the table; to be able to wait a few minutes to eat when hungry; to rely on internal cues to recognize fullness; to enjoy many different foods; to try new ones; to eat comfortably in places other than home. As Satter remarks, some of the adults reading her book may 'get the uneasy feeling that you haven't mastered all of these competencies yourself'. But there's still time.

Bibliography

Albala, Ken (2002), *Eating Right in the Renaissance*, Berkeley: University of California Press

Alberts, Hugo, Sandra Mulkens, Maud Smeets et al. (2010), 'Coping with Food Cravings. Investigating the Potential of a Mindfulness-based Intervention', *Appetite*, 55, 160–63

Alderman, Harold, John Hoddinott and Bill Kinsey (2006), 'Long term Consequences of Early Childhood Malnutrition', *Oxford Economic Papers*, 58, 450–74

Ali, E., R. Zachariah, A. Dahmane et al. (2013), 'Peanut-based Ready-to-use Therapeutic Food: Acceptability among Malnourished Children and Community Workers in Bangladesh', *Public Health Action*, 3, 128–35

Anderson, J.W., L. Grant, L. Gotthelf et al. (2007a), 'Weight Loss and Long-Term Follow-Up of Severely Obese Individuals Treated with an Intense Behavioral Program', *International Journal of Obesity*, 31, 488–93

Anderson, J.W., Shannon B. Conley and Amy S. Nicholas (2007b), 'One Hundred-pound Weight Losses with an Intensive Behavioral Program: Changes in Risk Factors in 118 Patients with Long-term Follow-up', *American Journal of Clinical Nutrition*, 86, 301–7

Añez, E., A. Remington, J. Wardle et al. (2012), 'The Impact of Instrumental Feeding on Children's Responses to Taste Exposure', *Journal of Human Nutrition and Dietetics*, 26, 415–20

Anliker, J.A., L. Bartoshuk, A.M. Ferris et al. (1991), 'Children's Food Preferences and Genetic Sensitivity to the Bitter Taste of 6-n-propylthiouracil (PROP)', *American Journal of Clinical Nutrition*, 54, 316–20

Appelhans, Bradley M., Molly E. Waring, Kristen Schneider et al. (2014), 'Food Preparation Supplies Predict Children's Family Meal and Home-prepared Dinner Consumption in Low-income Households', *Appetite*, 76, 1–8

Armstrong, Bridget and David M. Janicke (2012), 'Differentiating the Effects of Maternal and Peer Encouragement to Diet on Child Weight Control Attitudes and Behaviors', *Appetite*, 59, 723–9

Arnold, Carrie (2012), *Decoding Anorexia: How Breakthroughs in Science Offer Hope for Eating Disorders*, New York and London: Routledge

Asmaro, Devar, Fern Jaspers-Fayer and Valery Sramko (2012), 'Spatiotemporal Dynamics of the Hedonic Processing of Chocolate Images in Individuals with and without Trait Chocolate Craving', *Appetite*, 58, 790–99

Bailer, Ursula, Martina de Zwaan, Friedrich Leisch et al. (2004). 'Guided self-help versus cognitive-behavioral group therapy in the treatment of bulimia nervosa', *International Journal of Eating Disorders*, 35, 522–537.

Baron-Cohen, Simon, Tony Jaffa, Sarah Davies et al. (2013), 'Do Girls with Anorexia Nervosa have Elevated Autistic Traits?', *Molecular Autism*, 4, 24

Bartoshuk, Linda (2000), 'Comparing Sensory Experiences Among Individuals: Recent Psychophysical Advances Illuminate Genetic Variation in Taste Perception', *Chemical Senses*, 25, 443–60

Batsell, W. Robert, Alan S. Brown and Matthew E. Ansfield (2002), '"You Will Eat All of That!": A Retrospective Analysis of Forced Consumption Episodes', *Appetite*, 38, 211–19

Bauer, Katherine W., Melissa N. Laska, Jayne A. Fulkerson et al. (2011), 'Longitudinal and Secular Trends in Parental Encouragement for Healthy Eating, Physical Activity, and Dieting Throughout the Adolescent Years', *Journal of Adolescent Health*, 49, 306–11

Baumeister, Roy, Ellen Bratslavsky, Mark Muraven et al. (1998), 'Ego Depletion: Is the Active Self a Limited Resource?', *Journal of Personality and Psychology*, 74, 1252–65

Beauchamp, Gary and Julie A. Mennella (2011), 'Flavor Perception in Human Infants: Development and Functional Significance', *Digestion*, 83, 1–6

Beecher, Jonathan (1986), *Charles Fourier: The Visionary and his World*, Berkeley, London: University of California Press

Benelam, B. (2009), 'Satiation, Satiety, and their Effects on Eating Behaviour', *British Nutrition Foundation Bulletin*, 4, 126–73

Benjamin Murray, Stuart, Chris Thornton and Andrew Wallis (2013), 'Selective Eating in a 9-year-old Boy: Family Therapy as a First-line Treatment', *Clinical Child Psychology and Psychiatry*, 18, 270–75

Bentley, Amy (2006), 'Booming Baby Food: Infant Food and Feeding in Post-World War II America', *Michigan Historical Review*, 32, 63–87

Bentley, Margaret E., Heather Wasser and Hilary M. Creed-Kanashiro (2011), 'Responsive Feeding and Child Under-Nutrition in Low- and Middle-Income Countries', *Journal of Nutrition*, 141, 502–7

Berridge, Kent C. (2009), '"Liking" and "Wanting" Food Rewards: Brain Substrates and Roles in Eating Disorders', *Physiology and Behavior*, 97, 537–50

Birch, Leann L. (1998), 'Psychological Influences on the Childhood Diet', *Journal of Nutrition*, 128, 407S–10S

Birch, Leann L. (1999), 'Development of Food Preferences', *Annual Review of Nutrition*, 19, 41–62

Birch, Leann and Stephanie L. Anzman (2010), 'Learning to Eat in an Obesogenic Environment: A Developmental Systems Perspective on Child Obesity', *Child Development Perspectives*, 4, 138–43

Birch, Leann L. and Diane-Wolfe Marlin (1982), 'I Don't Like It; I Never Tried It: Effects of Exposure on Two-year-old Children's Food Preferences', *Appetite*, 3, 353–60

Bittman, Mark (2013), 'The Frankfurter Diaries', *New York Times*, 30 April

Blake, Anthony (2001), 'The Language of Flavour: Learning and Memory', in Harlan Walker (ed.), *Food and the Memory: Proceedings of the Oxford Symposium on Food and Cookery*, Totnes: Prospect Books

Blisset, Jacqueline, Caroline Meyer and Emma Haycraft (2006), 'Maternal and Paternal Controlling Feeding Practices with Male and Female Children', *Appetite*, 47, 212–19

Block, Robert W. and Nancy F. Krebs (2005), 'Failure to Thrive as a Manifestation of Child Neglect', *Pediatrics*, 116, 1234–7

Blumenthal, Heston (2009), *The Fat Duck Cookbook*, London: Bloomsbury

Boorstin, Sharon (2001), 'Kids' Menus: Keep 'Em Happy', *Restaurant Hospitality*, 85, 95

Bostic, Jeff Q., Anna C. Muriel, Sabine Hack et al. (1997), 'Anorexia Nervosa in a 7-Year-Old Girl', *Developmental and Behavioral Pediatrics*, 18, 331–3

Botz-Bornstein, Thorsten and Noreen Abdullah-Khan (2014), *The Veil in Kuwait: Gender, Fashion, Identity*, London: Palgrave

Bourdieu, Pierre (1986), *Distinction: A Social Critique of the Judgment of Taste*, trans. Richard Nice, London: Routledge, Kegan & Paul

Bowen, Deborah, Carolyn Ehret, Margaret Pederson et al. (2002), 'Results of an Adjunct Dietary Intervention Program in the Women's Health Initiative', *Journal of the American Dietetic Association*, 102, 1631–37

Breen, Fiona M., Robert Plomin and Jane Wardle (2006), 'Heritability of Food Preferences in Young Children', *Physiology and Behavior*, 88, 443–7

Brillat-Savarin, Jean Anthelme (2009), *The Physiology of Taste*, trans. M.F.K. Fisher with introduction by Bill Buford, London: Everyman

Brown, Harriet (2009), *Brave Girl Eating*, New York: William Morrow

Brožek, J. Josef (1953), 'Semistarvation and Nutritional Rehabilitation: A Qualitative Case Study with Emphasis on Behavior', *American Journal of Clinical Nutrition*, 1, 107–18

Bruch, Hilde (1974), *Eating Disorders: Obesity, Anorexia Nervosa and the Person Within*, London: Routledge

Bruch, Hilde (1978), *The Golden Cage: The Enigma of Anorexia Nervosa*, London: Open Books

Bryant-Waugh, Rachel (2013), 'Avoidant Restrictive Food Intake Disorder: An Illustrative Case Example', *International Journal of Eating Disorders*, 46, 420–23

Bryant-Waugh, Rachel, Laura Markham, Richard Kreipe et al. (2010), 'Feeding and Eating Disorders in Childhood', *International Journal of Eating Disorders*, 43, 98–111

Burd, Carlye, Araliya Senerat, Earkle Chambers et al. (2013), 'PROP Taster Status Interacts with the Built Environment to Influence Children's Food Acceptance and Body Weight Status', *Obesity*, 21, 786–94

Carafoli, John F. (2001), 'Amarcord: The Flavor of Buried Memories', in Harlan Walker (ed.), *Food and the Memory: Proceedings of the Oxford Symposium on Food and Cookery*, Totnes: Prospect Books

Carlson, Anton J. (1993), 'Contributions to the Physiology of the Stomach. – II. The Relation between the Contractions of the Empty Stomach and the Sensation of Hunger', *Obesity Research*, vol. 1 no. 6, 501–9

Carnell, S., L. Cooke, R. Cheng et al. (2011), 'Parental Feeding Behaviours and Motivations. A Quantitative Study in Mothers of UK Pre-schoolers', *Appetite*, 57, 665–73

Carruth, Betty Ruth, Paula J. Ziegler, Anne Gordon et al. (2004), 'Prevalence of Picky Eaters Among Infants and Toddlers and Their Caregivers' Decisions about Offering a New Food', *Journal of the American Dietetic Association*, 104, S57–S64

Castonguay, Jessica, Dale Kunkel, Paul Wright et al. (2013), 'Healthy Characters? An Investigation of Marketing Practices in Children's Food Advertising', *Journal of Nutrition Education and Behavior*, 45, 571–7

Castro, D.C. and K.C. Berridge (2014), 'Advances in the Neurobiological Bases for Food "Liking" versus "Wanting"', *Physiology and Behavior*, 136, 22–30

Catanzaro, Diane, Emily C. Chesbro and Andrew J. Velkey (2013), 'Relationship between Food Preferences and PROP Taster Status of College Students', *Appetite*, 68, 124–31

Cathro, Jo and Moira Hilliam (1994), *Children's Eating Habits in Europe*, Leatherhead, Surrey: Leatherhead Food Research

CDC, Centers for Disease Control and Prevention (2010), 'State-Specific Trends in Fruit and Vegetable Consumption Among Adults – United States, 2000–2009', *Morbidity and Mortality Weekly Report*, September

Chapman, Gwen and Heather Maclean (1993), '"Junk Food" and "Healthy Food": Meanings of Food in Adolescent Women's Culture', *Society for Nutrition Education*, 25, 108–13

Chapman, Katarzyna and Jane Ogden (2010), 'The Prevalence of Mechanisms of Dietary Change in a Community Sample', *Appetite*, 55, 447–53

Clark, Georgiana C. (1874), *Economical Cookery*, London: Simpkin, Marshall & Co.

Clifton, Claire and Colin Spencer (1993), *The Faber Book of Food*, London: Faber

Coates, Anne (1996), *Your Only Child*, London: Bloomsbury

Cole, S.Z. and J.S. Lanham (2011), 'Failure to Thrive: An Update', *American Family Physician*, 83, 829–34

Collingham, Lizzie (2011), *The Taste of War: World War II and the Battle for Food*, London: Penguin

Conley, Dalton and Rebecca Glauber (2007), 'Gender, Body Mass and Socioeconomic Status: New Evidence from the PSID', *Advances in Health Economics*, 17, 253–75

Cooke, Lucy, Susan Carnell and Jane Wardle (2006), 'Food Neophobia and Mealtime Food Consumption in 4–5 year old Children', *International Journal of Behavioral Nutrition and Physical Activity*, 3

Cooke, L.J., L.C. Chambers, E.V. Añez et al. (2011), 'Eating for Pleasure or Profit: The Effect of Incentives on Children's Enjoyment of Vegetables', *Psychological Science*, 22, 190–96

Cornwell, T. Bettina and Anna R. McAlister (2011), 'Alternative Thinking about Starting Points of Obesity. Development of Child Taste Preferences', *Appetite*, 56, 428–39

Cortes, D.E., A. Millan-Ferro, K. Schneider et al. (2013), 'Food Purchasing Selection among Low-income, Spanish-speaking Latinos', *American Journal of Preventive Medicine*, 44, S267–73

Coulthard, Helen, Gillian Harris and Anna Fogel (2014), 'Exposure to Vegetable Variety in Infants Weaned at Different Ages', *Appetite*, 78C, 89–94

Cowart, B.J. (1981), 'Development of Taste Perception in Humans: Sensitivity and Preference throughout the Life Span', *Psychological Bulletin*, 90, 43–73

Crowley, Ralph H. (1909), *The Hygiene of School Life*, London: Methuen

Cruwys, Tegan, Kirsten E. Bevelander and Roel C.J. Hermans (2015), 'Social Modeling of Eating: A Review of When and Why Social Influence Affects Food Intake and Choice', *Appetite*, 86, 3–18

Culpeper, Nicholas (1662), *Culpeper's Directory for Midwives*, London: Peter Cole

Cutts, Diana Becker, Alan F. Meyers, Maureen Black et al. (2011), 'US Housing Insecurity and the Health of Very Young Children', *American Journal of Public Health*, 101, 1508–14

Cwiertka, Katarzyna J. (2006), *Modern Japanese Cuisine*, London: Reaktion

Cwiertka, Katarzyna J. (2012), *Cuisine, Colonialism and Cold War: Food in Twentieth-century Korea*, London: Reaktion

Dalton, P., N. Doolittle, H. Nagata et al. (2000), 'The Merging of the Senses: Integration of Subthreshold Taste and Smell', *Nature Neuroscience*, 3, 431–2

David, Elizabeth (2000), *Is There a Nutmeg in the House?*, compiled by Jill Norman, London: Michael Joseph

Davis, Clara M. (1928), 'Self Selection of Diet by Newly Weaned Infants', *American Journal of Diseases of Children*, 36, 651–79

Davis, Clara M. (1939), 'Results of the Self-Selection of Diets by Young Children', *Canadian Medical Association Journal*, 41, 257–61

De Graaf, Cees, Wendy Blom, Paul Smeets et al. (2004), 'Biomarkers of Satiation and Satiety', *American Journal of Clinical Nutrition*, 79, 946–61

Delaney, Charlotte B., Kamryn T. Eddy and Andrea Hartmann (2014), 'Pica and Rumination Behavior among Individuals Seeking Treatment for Eating Disorders or Obesity', *International Journal of Eating Disorders*, 48, 238–48

De Leeuw, Rebecca, Harriëtte M. Snoek, Jan F.J. van Leeuwe et al. (2007), 'Similarities and Reciprocal Influences in Eating Behavior within Sibling Pairs: A Longitudinal Study', *Eating Behaviors*, 8, 464–73

De Sa, Joia et al. (2013), 'Identifying Priorities to Improve Maternal and Child Nutrition among the Khmu Ethnic Group, Laos: A Formative Study', *Maternal and Child Nutrition*, 9, 456–66

Devine, Carol M. (2005), 'A Life Course Perspective: Understanding Food Choices in Time, Social Location, and History', *Journal of Nutrition Education and Behavior*, 37, 121–8

Dinehart, M.E., J.E. Hayes, L.M. Bartoshuk et al. (2006), 'Bitter Taste Markers Explain Variability in Vegetable Sweetness, Bitterness and Intake', *Physiology and Behavior*, 87, 304–13

Dovey, Terence M., Paul A. Staples, E. Leigh Gibson et al. (2008), 'Food Neophobia and "Picky/Fussy" Eating in Children: A Review', *Appetite*, 50, 181–93

Drewnowski, Adam (1997), 'Taste Preferences and Food Intake', *Annual Review of Nutrition*, 17, 237–53

Drewnowski, Adam, Julie Mennella, Susan Johnson et al. (2012), 'Sweetness and Food Preference', *Journal of Nutrition*, 142, 1142S–8S

Druckerman, Pamela (2013), *French Children Don't Throw Food*, London: Black Swan

Duffy, Valerie B., Andrew C. Davidson, Judith R. Kidd et al. (2004), 'Bitter Receptor Gene (TAS2R38), PROP Bitterness and Alcohol Intake', *Alcoholism Clinical and Experimental Research*, 28(11): 1629–37

Duncker, Karl (1938), 'Experimental Modification of Children's Food Preferences Through Social Suggestion', *Journal of Abnormal and Social Psychology*, 33, 489–507

Duncker, Karl (1939), 'The Influence of Past Experience Upon Perceptual Properties', *American Journal of Psychology*, 52, 255–65

Duncker, Karl (1941), 'On Pleasure, Emotion and Striving', *Philosophy and Phenomenological Research*, 1, 391–430

Dutton, Thomas (1906), *The Mother's Guide to the Feeding and Rearing of Children*, London: Henry Kimpton, 3rd edn

Eftekhari, M.H., H. Mozaffari-Khosravi and F. Shidfar (2009), 'The Relationship between BMI and Iron Status in Iron-Deficient Adolescent Iranian Girls' *Public Health Nutrition*, 12, 2377–81.

Elfhag, K. and S. Rössner (2005), 'Who Succeeds in Maintaining Weight Loss? A Conceptual Review of Factors Associated with Weight Loss Maintenance and Regain', *Obesity Reviews*, 6, 67–85

Elliott, Charlene (2008), 'Marketing Fun Foods: A Profile and Analysis of Supermarket Food Messages Targeted at Children', *Canadian Public Policy*, 34, 259–73

Ernsperger, Lori and Tania Stegen-Hanson (2004), *Just Take a Bite: Easy, Effective Answers to Food Aversions and Eating Challenges*, Arlington, Texas: Future Horizons

Evers, Catherine, Marieke Adriaanse et al. (2013), 'Good Mood Food: Positive Emotion as a Neglected Trigger for Food Intake', *Appetite*, 68, 1–7

Faith, Myles S., Robert I. Berkowitz et al. (2006), 'Eating in the Absence of Hunger: A Genetic Marker for Obesity in Prepubertal Boys?', *Obesity*, vol. 14, no. 1, 131–8

Faith, M.S., A. Pietrobelli, M. Heo et al. (2012), 'A Twin Study of Self-regulatory Eating in Early Childhood: Estimates of Genetic and Environmental Influence, and Measurement Considerations', *International Journal of Obesity*, 36, 931–7

Fallani, M., D. Young, J. Scott et al. (2010), 'Intestinal Microbiota of 6-week-old Infants across Europe: Geographic Influence beyond Delivery Mode, Breast-feeding, and Antibiotics', *Journal of Pediatric Gastroenterology*, 51, 77–84

Fallon, April, Paul Rozin and Patricia Pliner (1984), 'The Child's Conception of Food: The Development of Food Rejections with Special Reference to Disgust and Contamination Sensitivity', *Child Development*, 55, 566–75

Farris, Alisha, Sarah Misyak et al. (2014), 'Nutritional Comparison of Packed and School Lunches in Pre-Kindergarten and Kindergarten Children Following the Implementation of the 2012–2013 National School Lunch Program Standards', *Journal of Nutrition Education and Behavior*, 46, 621–6

Feeney, Emma L., Sinead A. O'Brien, Amalia G.M. Scannell et al. (2014), 'Genetic and Environmental Influences on Liking and Reported Intakes of Vegetables in Irish Children', *Food Quality and Preference*, 32, 253–63

Ficker, Victor B. and Herbert S. Graves (eds) (1971), *Deprivation in America*, Beverly Hills: Glencoe Press

Fildes, A., C.H. van Jaarsveld, C.H. Llewellyn et al. (2014), 'Nature and Nurture in Children's Food Preferences', *American Journal of Clinical Nutrition*, 99, 911–17

Fisher, Jennifer Orlet and Leann L. Birch (2002), 'Eating in the Absence of Hunger and Overweight in Girls from 5 to 7 Years of Age', *American Journal of Clinical Nutrition*, 76, 226–31

Fisher, Jennifer O., Guowen Cai et al. (2007), 'Heritability of Hyperphagic Eating Behaviour and Appetite-related Hormones among Hispanic Children', *Obesity*, vol. 15, no. 6, 1484–95

Fisher, Martin M., David S. Rosen and Rollyn M. Ornstein (2014), 'Characteristics of Avoidant/Restrictive Food Intake Disorder in Children and Adolescents: A "New Disorder" in DSM-5', *Journal of Adolescent Health*, 1–4

Fong, Vanessa (2004), *Only Hope: Coming of Age Under China's One-Child Policy*, Stanford: Stanford University Press

French, Paul and Matthew Crabbe (2010), *Fat China: How Expanding Waistlines are Changing a Nation*, London: Anthem Press

Fuhrer, Dagmar, Stefan Zysset and Michael Stumvoll (2008), 'Brain Activity in Hunger and Satiety: An Exploratory Visually Stimulated fMRI Study', *Obesity*, vol. 16, no. 5, 945–50

Fulkerson, J.A., J. Strauss, D. Neumark-Sztainer et al. (2007), 'Correlates of Psychosocial Well-being among Overweight Adolescents: The Role of the Family', *Journal of Consulting and Clinical Psychology*, 75, 181–6

Galloway, Amy T., Laura M. Fiorito, Lori A. Francis et al. (2006), '"Finish Your Soup": Counterproductive Effects of Pressuring Children to Eat on Intake and Affect', *Appetite*, 46, 318–23

Garcia, Olga P., Kurt Long and Jorge L. Rosado (2009), 'Impact of Micronutrient Deficiencies on Obesity', *Nutrition Reviews*, 67, 559–72

Geier, Andrew B. and Paul Rozin (2003), 'Weighing Discomfort in College Age American Females: Incidence and Causes', *Appetite*, 51, 173–7

Goh, Esther C.L. (2009), 'Grandparents as Childcare Providers: An In-depth Analysis of the Case of Xiamen, China', *Journal of Aging Studies*, 23, 60–68

Gold, Rich (1993), 'Art in the Age of Ubiquitous Computing', *American Art*, 7, 2–11

Goldberg, Joan (1990), '"Wisdom of the Body" May Determine Food Cravings: Nutrition Theory Points to Link Between Obsessive Thoughts for Certain Foods and Biological Needs', *Los Angeles Times*, 23 November

Gonzalez, Kristina M., Catherine Peo, Tod Livdahl et al. (2008), 'Experience-Induced Changes in Sugar Taste Discrimination', *Chemical Senses*, 33, 173–9

Gopnik, Adam (2011), *The Table Comes First: Family, France and the Meaning of Food*, London: Quercus

Groves, Angela (2002), *Children's Food: Market Forces and Industry Responses*, Watford: IGD

Gugusheff, J.R., M. Vithayathil, Z.Y. Ong et al. (2013), 'The Effects of Prenatal Exposure to a "Junk Food" Diet on Offspring Food Preferences and Fat Deposition can be Mitigated by Improved

Nutrition during Lactation', *Journal of Developmental Origins of Health and Disease*, 4, 348–57

Haase, Lori, Erin Green and Claire Murphy (2011), 'Males and Females Show Differential Brain Activation to Taste When Hungry', *Appetite*, 57, 421–34

Hales, C. Nicholas and David J.P. Barker (2001), 'The Thrifty Phenotype Hypothesis', *British Medical Bulletin*, 60, 5–20

Haller, R., C. Rummel, S. Henneberg et al. (1999), 'The Influence of Early Experience with Vanillin on Food Preference Later in Life', *Chemical Senses*, 24, 465–7

Hammons, Amber J. and Barbara H. Fiese (2011), 'Is Frequency of Shared Family Meals Related to the Nutritional Health of Children and Adolescents?', *Pediatrics*, 127, e1565–74

Hardyment, Christina, (1995), *Perfect Parents: baby-care advice past and present*, Oxford: Oxford University Press

Hare, Caspar (2010), 'Take the Sugar', *Analysis*, 70, 237–47

Harris, Gillian (2008), 'Development of Taste and Food Preferences in Children', *Clinical Nutrition and Metabolic Care*, 11, 315–19

Havermans, Remco C. (2011), '"You Say it's Liking, I Say it's Wanting . . .". On the Difficulty of Disentangling Food Reward in Man', *Appetite*, 57, 286–94

Hay, Phillipa J. and Perminder Sachdev (2011), 'Brain Dysfunction in Anorexia Nervosa: Cause or Consequence of Under-nutrition?', *Current Opinion in Psychiatry*, 24, 251–6

He, Meizi and Anita Evans (2007), 'Are Parents Aware that their Children are Overweight or Obese? Do They Care?', *Canadian Family Physician*, 53, 1493–9

Hecht, Charles (ed.) (1912), *Our Children's Health at Home and at School Being the Report of a Conference on Diet and Hygiene in Public Secondary & Private Schools held at the Guildhall*, London, 13 May, London: National Food Reform Association

Hecht, Charles (ed.) (1913), *Rearing an Imperial Race: Containing a Full Report of the Second Guildhall Conference on Diet, Cookery and Hygiene*, London: National Food Reform Association

Hendy, Helen M. and Keith E. Williams (2012), 'Mothers' Feeding Practices for Children 3–10 Years of Age and their Associations with Child Demographics', *Appetite*, 58, 710–16

Henry, Diana, (2014), *A Change of Appetite: Where delicious meets healthy*, London: Mitchell Beazley

Hercberg, Serge, Paul Preziosi and Pilar Galan (2001), 'Iron Deficiency in Europe', *Public Health Nutrition*, 4, 537–45

Herman, C. Peter, Nicola E. Fitzgerald and Janet Polivy (2003), 'The Influence of Social Norms on Hunger Ratings and Eating', *Appetite*, 41, 15–20

Herrin, Marcia and Marcia Larkin (2013), *Nutrition Counseling in the Treatment of Eating Disorders*, New York: Brunner-Routledge

Herzog, David B., David J. Dorer and Pamela Keel (1999), 'Recovery and Relapse in Anorexia and Bulimia Nervosa: A 7.5-Year Follow-up Study', *Journal of the American Academy of Child and Adolescent Psychiatry*, 38, 829–37

Hilliam, Moira (1996), *European Market Opportunities in Children's Food and Drink, Winning Children as Customers*, FT Management Reports, London: Pearson

Hirschmann, J.R. and L. Zaphiropoulos (1985), *Solving Your Child's Eating Problems*, New York: Fawcett Columbine

Hoefling, Atilla, Katja Likowski, Michael Hafner et al. (2009), 'When Hunger Finds no Fault with Moldy Corn: Food Deprivation Reduces Food-related Disgust', *Emotion*, 9, 50–58

Hoek, Hans Wijbrand and Daphne van Hoeken (2003), 'Review of the Prevalence and Incidence of Eating Disorders', *International Journal of Eating Disorders*, 34, 383–96

Hoerr, S.L., S.O. Hughes, J.O. Fisher et al. (2009), 'Associations among Parental Feeding Styles and Children's Food Intake in Families with Limited Incomes', *International Journal of Behavioral Nutrition and Physical Activity*, 6, 55–62

Holsten, Joanna E., Janet A. Deatrick, Shiriki Kumanyika et al. (2011), 'Children's Food Choice Process in the Home Environment. A Qualitative Descriptive Study', *Appetite*, 58, 64–73

Holt, L. Emmet (1923), *The Care and Feeding of Children*, New York: D. Appleton, 8th edn [first edn 1894]

Hormes, Julia and Paul Rozin (2009), 'Perimenstrual Chocolate Craving: What Happens after Menopause?', *Appetite*, 53, 256–9

Howard, Natasha J., Graeme J. Hugo, Anne Taylor et al. (2008), 'On Perception of Weight: Socioeconomic and Sociocultural Explanations', *Obesity Research and Clinical Practice*, 2, 125–31

Huang, Shirley H., Elizabeth P. Parks, Shiriki K. Kumanyika et al. (2012), 'Child-feeding Practices among Chinese-American and Non-Hispanic White Caregivers', *Appetite*, 58, 922–7

Hubble, Helen and Florence G. Blake (1944), 'Feeding Children in Wartime', *American Journal of Nursing*, 44, 445–8

Hughes, Georgina, Kate M. Bennett and Marion M. Hetherington (2004), 'Old and Alone: Barriers to Healthy Eating in Older Men Living on their Own', *Appetite*, 43, 269–73

Humble, Nicola (2010), *Cake: A Global History*, London: Reaktion

Ishige, Naomichi (2001), *The History and Culture of Japanese Food*, London: Kegan Paul

Itard, Jean-Marc-Gaspard (1932), *The Wild Boy of Aveyron*, translated by George and Muriel Humphrey, New York, London: The Century Company

Itoh, Makiko (2011), *The Just Bento Cookbook*, New York: Kodansha USA

Jain, Anjali, Susan N. Sherman, Leigh A. Chamberlin (2001), 'Why Don't Low-Income Mothers Worry About their Preschoolers Being Overweight?', *Pediatrics*, 107, 1138

Jelliffe, Derrick B. (1962), 'Culture, Social Change and Infant Feeding', *American Journal of Clinical Nutrition*, 10, 19–45

Jennings, Lisa (2009), 'Survey says Kids Menus Need Healthier Offerings', *Nation's Restaurant News*, 17 August

Jerzsa-Latta, Margaret, Magdelena Krondl and Patricia Coleman (1990), 'Use and Perceived Attributes of Cruciferous Vegetables in Terms of Genetically-Mediated Taste Sensitivity', *Appetite*, 1990, 15, 127–34

Jingxiong, Jiang, Urban Rosenqvist and Wang Huishan (2007), 'Influence of Grandparents on Eating Behaviors of Young Children in Chinese Three-generation Families', *Appetite*, 48, 377–83

Johnson, S.L. (2000), 'Improving Preschooler's Self-Regulation of Energy Intake', *Pediatrics*, 106, 1429–35

Kaminski, Linda Clancy, Susan Henderson and Adam Drewnowski (2000), 'Young Women's Food Preferences and Taste Responsiveness to 6-n-propylthiouracil (PROP)', *Physiology and Behavior*, 68, 691–7

Karmel, Annabel (1991), *The Complete Baby and Toddler Meal Planner: Over 200 Quick, Easy and Healthy Recipes*, London: Ebury

Katz, David (2014), 'Knowing What to Eat, Refusing to Swallow It', *Huffington Post*, posted 7 February

Katz, David and S. Meller (2014), 'Can We Say What Diet is Best for Health?', *Annual Review of Public Health*, 35, 83–103

Kauer, Jane, Marcia Pelchat, Paul Rozin et al. (2015), 'Adult Picky Eating. Phenomenology, Taste Sensitivity and Psychological Correlates', *Appetite*, 90, 219–28

Kawash, Samira (2013), *Candy: A Century of Panic and Pleasure*, London: Faber & Faber

Kayman, Susan, William Bruvold and Judith S. Stern (1990), 'Maintenance and Relapse after Weight Loss in Women: Behavioral Aspects', *American Journal of Clinical Nutrition*, 52, 800–807

Keller, Heather H., Margaret Hedley, Teresa Hadley et al. (2005), 'Food Workshops, Nutrition Education and Older Adults', *Journal of Nutrition for the Elderly*, 24, 5–23

Keys, A., J. Brožek, A. Henschel et al. (1950), *The Biology of Human Starvation*, Oxford, England: University of Minnesota Press, 2 vols

Kimura, Atsushi, Yuji Wada, Akio Asakawa et al. (2012), 'Dish Influences Implicit Gender-based Food Stereotypes among Young', *Appetite*, 58, 940–45

Kimura, Atsushi, Yuji Wada and Sho-ichi Goto (2009), 'Implicit Gender-based Food Stereotypes. Semantic Priming Experiments on Young Japanese', *Appetite*, 52, 521–8

Kissileff, Harry R., Julie C. Carretta, Allan Geliebter et al. (2003), 'Cholecystokinin and Stomach Distension Combine to Reduce Food Intake in Humans', *American Journal of Physiology – Regulatory, Integrative and Comparative Physiology*, 285, R992–R998

Klump, Kelly L. (2013), 'Puberty as a Critical Risk Period for Eating Disorders: A Review of Human and Animal Studies', *Hormones and Behavior*, 64, 399–410

Koistinen, Aila and Leena Ruhanen (eds) (2009), 'To the World of Food with the Aid of the Senses: The Sapere Method as a Support for Children's Food and Nutrition Education in Daycare Centres', Jyväskylä: Sitra (available on the Finnish Sapere website, accessed August 2014)

Komatsu, Sakura (2008), 'Rice and Sushi Craving: A Preliminary Study of Food Craving among Japanese Females', *Appetite*, 50, 353–8

Köster, E.P. (2003), 'The Psychology of Food Choices: Some Often Encountered Fallacies', *Food Quality and Preference*, 14, 359–73

Köster, E.P. (2009), 'Diversity in the Determinants of Food Choice: A Psychological Perspective', *Food Quality and Preference*, 20, 70–82

Köster, E.P. and J. Mojet (2007), 'Boredom and the Reason Why Some New Food Products Fail', in H. MacFie (ed.), *Consumer-Led Food Product Development*, Cambridge: Woodhead

Köster, E.P., C. Rummel, C. Kornelson et al. (2001), 'Stability and Change in Food Liking: Food Preferences in the Two Germanys after the Reunification', in M. Roth (ed.), *Flavour 2000: Perception, Release, Evaluation, Formation, Acceptance, Nutrition and Health*, Bergholz-Rehbrücke, Germany: Rothe

Kotler, Lisa A., Patricia Cohen, Mark Davies et al. (2001), 'Longitudinal Relationships Between Childhood, Adolescent, and Adult Eating Disorders', *Journal of the American Academy of Child and Adolescent Psychiatry*, vol. 40, no. 12, 1434–40

Kovacs, Eva M.R., M.S. Westerterp-Plantenga, W.H.M. Saris et al. (2002), 'Associations between Spontaneous Meal Initiations and Blood Glucose Dynamics in Overweight Men in Negative Energy Balance', *British Journal of Nutrition*, 87, 39–45

Kristensen, S.T., L. Holm, A. Raben et al. (2002), 'Achieving "Proper" Satiety in Different Social Contexts – Qualitative Interpretations from a Cross-Disciplinary Project', *Appetite*, 39, 207–15

Kuchler, F. and J.N. Variyam (2003), 'Mistakes Were Made: Misperception as a Barrier to Reducing Overweight', *International Journal of Obesity and Related Metabolic Disorders*, 7, 856–61

Kuijer, Roeline G. and Jessica A. Boyce (2014), 'Chocolate Cake. Guilt or Celebration? Associations with Healthy Eating', *Appetite*, 74, 48–54

Kushner, Barak (2012), *Slurp! A Social and Culinary History of Ramen – Japan's Favorite Noodle Soup*, Leiden, Boston: Global Oriental

Lask, Bryan and Rachel Bryant-Waugh (eds) (2013), *Eating Disorders in Childhood and Adolescence*, London: Routledge

Laurier, Eric and Sally Wiggins (2011), 'Finishing the Family Meal. The Interactional Organization of Satiety', *Appetite*, 56, 53–64

Laybourn, Ann (1994), *The Only Child: Myths and Reality*, Edinburgh: HMSO

Lehmann, Gilly (2003), *The British Housewife: Cookery Books, Cooking and Society in 18th-Century Britain*, Totnes: Prospect Books

Leigh Gibson, Edward (2001), 'Learning in the Development of Food Craving', in Marion Hetherington (ed.), *Food Cravings and Addiction*, Leatherhead: Leatherhead Food Publishing

Levin, Kate A. and Joanna Kirby (2012), 'Irregular Breakfast Consumption in Adolescence and the Family Environment: Underlying Causes by Family Structure', *Appetite*, 59, 63–70

Levin Pelchat, Marcia and Fritz Blank (2001), 'A Scientific Approach to Flavours and Olfactory Memory', in Harlan Walker (ed.), *Food and the Memory: Proceedings of the Oxford Symposium on Food and Cookery*, Totnes: Prospect Books

Levin Pelchat, Marcia, Andrea Johnson, Robin Chan et al. (2004), 'Images of Desire: Food-craving Activation during fMRI', *Neuroimage*, 23, 1486–93

Lévy, C.M., A. MacRae and E.P. Köster (2006), 'Perceived Stimulus Complexity and Food Preference Development', *Acta Psychologica*, 123, 394–413

Lim, Stephen S., Theo Vos, Abraham D. Flaxman et al. (2012), 'A Comparative Risk Assessment of Burden of Disease and Injury Attributable to 67 Risk Factors and Risk Factor Clusters in 21 Regions, 1990–2010: A Systematic Analysis for the Global Burden of Disease Study 2010', *Lancet*, 380, 2224–60

Llewellyn, Clare H., Cornelia H.M. van Jaarsveld, Laura Johnson et al. (2010), 'Nature and Nurture in Infant Appetite: Analysis of the Gemini Twin Birth Cohort', *American Journal of Clinical Nutrition*, 91, 1172–9

Lobstein, Tim (1988), *Children's Food: The Good, the Bad and the Useless*, London: Unwin Paperbacks

Lock, James and Daniel Le Grange (2005), *Help Your Teenager Beat an Eating Disorder*, London: The Guilford Press

Lucas, Anna, Esther Murray and Sanjay Kinra (2013), 'Health Beliefs of UK South Asians Related to Lifestyle Diseases: A Review of Qualitative Literature', *Journal of Obesity*, 1–13

Lustig, Robert (2014), *Fat Chance: The Hidden Truth about Sugar, Obesity and Disease*, London: Fourth Estate

Lustig, Robert, Laura Schmidt and Claire D. Brindis (2012), 'The Toxic Truth About Sugar', *Nature*, 482, 27–9

McMillan, Margaret (1909), *London's Children: How to Feed Them and How Not to Feed Them*, London: Independent Labour Party

Madise, Nyovani J., Zoe Matthews and Barrie Margetts (1999), 'Heterogeneity of Child Nutritional Status between Households: A Comparison of Six Sub-Saharan African Countries', *Population Studies*, vol. 53, no. 3, 331–43

Maier, Andrea, Claire Chabanet, Benoist Schaal et al. (2007), 'Effects of Repeated Exposure on Acceptance of Initially Disliked Vegetables in 7-month-old Infants', *Food Quality and Preference*, 18, 1023–32

Malnic, Bettina, Junzo Hirono, Takaaki Sato et al. (1999), 'Combinatorial Receptor Codes for Odors', *Cell*, 96, 713–23

Marshall, C.F. (1895), 'A Fatal Case of Anorexia Nervosa', *Lancet*, 19 January

Martens, Lydia (1997), 'Gender and the Eating Out Experience', *British Food Journal*, 99, 20–26

Mattes, Richard (1990), 'Hunger Ratings are not a Valid Proxy Measure of Reported Food Intake in Humans', *Appetite*, 15, 103–13

Mattes, Richard (1997), 'The Taste for Salt in Humans', *American Journal of Clinical Nutrition*, 65, 692S–7S

Mattes, Richard (2005), 'Soup and Satiety', *Physiology and Behavior*, 83, 739–47

Mattes, Richard (2010), 'Hunger and Thirst: Issues in Measurement and Prediction of Eating and Drinking', *Physiology and Behavior*, 100, 22–32

Mead, Margaret (ed.) (1937), *Cooperation and Competition Among Primitive Peoples*, New York: McGraw-Hill

Mead, Margaret (1943), 'The Factor of Food Habits', *Annals of the American Academy of Political and Social Science*, 225, 136–41

Meiselman, Herbert L. (2006), 'The Role of Context in Food Choice, Food Acceptance and Food Consumption', in Richard Shepherd and Monique Raats (eds), *The Psychology of Food Choice*, Wallingford: CABI, 179–201

Meiselman, Herbert L. and H.J.H. MacFie (1996), *Food Choice, Acceptance and Consumption*, London, New York, Tokyo: Blackie

Mendelson, Charlotte (2013), 'Forty Words of Love in Hungarian', *Guardian*, 10 August

Mennell, Stephen (1985), *All Manners of Food: Eating and Taste in England and France from the Middle Ages to the Present Time*, Oxford: Basil Blackwell

Mennella, J.A. and G.K. Beauchamp (1991), 'Maternal Diet Alters the Sensory Qualities of Human Milk and the Nursling's Behavior', *Pediatrics*, 88, 737–44

Mennella, J.A. and G.K. Beauchamp (1993), 'The Effects of Repeated Exposure to Garlic-flavored Milk on the Nursling's Behavior', *Pediatric Research*, 34, 805–8

Mennella, J.A., Anthony Johnson and Gary Beauchamp (1995), 'Garlic Ingestion by Pregnant Women Alters the Odor of Amniotic Fluid', *Chemical Senses*, 20, 207–9

Mennella, Julie, Yanina Pepino and Danielle Reed (2005), 'Genetic and Environmental Determinants of Bitter Perception and Sweet Preferences', *Pediatrics*, vol. 115, no. 2, February

Miller, William R. and Stephen Rollnick (2013), *Motivational Interviewing: Helping People Change*, New York: Guilford Press

Mirch, Margaret, Jennifer R. McDuffie and Susan Z. Yanovski (2006), 'Effects of Binge Eating on Satiation, Satiety, and Energy Intake of Overweight Children', *American Journal of Clinical Nutrition*, 84, 732–8

Mitrany, Edith (1992), 'Atypical Eating Disorders', *Journal of Adolescent Health*, 13, 400–402

Mojet, J. and E.P. Köster (2005), 'Sensory Memory and Food Texture', *Food Quality and Preference*, 16, 251–66

Mojet, J. and E.P. Köster (2006), 'Theories of Food Choice Development', in L. Frewer and H. van Trijp (eds), *Understanding Consumers of Food Products*, Cambridge: Woodhead Publishing

Møller, Per, Jos Mojet, Egon Peter Köster (2007), 'Incidental and Intentional Flavor Memory in Young and Older Subjects', *Chemical Senses*, 32, 557–67

Monello, Lenore F. and Jean Mayer (1967), 'Hunger and Satiety Sensations in Men, Women, Boys and Girls', *American Journal of Clinical Nutrition*, vol. 20, no. 3, 253–61

Moore, Anna (2011), 'Life After an Eating Disorder', *Daily Telegraph*, 27 November

Morbidity and Mortality Weekly Report 1994, 'Daily dietary fat and total food-energy intakes – third National Health Nutrition Examination Survey, Phase I, 1988–91', MMWR, 43, 116–23

Moss, Michael (2014), *Salt, Sugar, Fat: How the Food Giants Hooked Us*, London: W.H. Allen

Musaiger, A.O., Mariam Al-Mannai, Reema Tayyem et al. (2012), 'Prevalence of Overweight and Obesity among Adolescents in Seven Arab Countries: A Cross-Cultural Study', *Journal of Obesity*, 1–5

Musaiger, A.O., Mariam Al-Mannai, Reema Tayyem et al. (2013), 'Risk of Disordered Eating Attitudes among Adolescents in Seven Arab Countries', *Appetite*, 60, 162–7

Mustonen, Sari and Hely Tuorila (2010), 'Sensory Education Decreases Food Neophobia Score and Encourages Trying Unfamiliar Foods in 8–12-year-old Children', *Food Quality and Preference*, 21, 353–60

Naser Al-Isa, A., J. Campbell and E. Desapriya (2013), 'Factors Associated with Overweight and Obesity among Kuwaiti Men', *Asia Pacific Journal of Public Health*, 25, 63

Natow, Annette B. and Jo-Ann Heslin (1982), 'Nutrition Education in Later Years', *Journal of Nutrition for the Elderly*, 1, 101–20

Nelson, M. (1996), 'Anaemia in Adolescent Girls: Effects on Cognitive Function and Activity', *Proceedings of the Nutrition Society*, 55, 359–67

Nestle, Marion (2007), *What to Eat*, New York: North Point Press

Nestle, Marion, Rena Wing, Leann Birch et al. (1998), 'Behavioural and Social Influences on Food Choices', *Nutrition Reviews*, vol. 56, no. 5, S50–S74

Neumark-Sztainer, Dianne, Katherine W. Bauer, Sarah Friend et al. (2010), 'Family Weight Talk and Dieting: How Much Do They Matter for Body Dissatisfaction and Disordered Eating Behaviors in Adolescent Girls?', *Journal of Adolescent Health*, 47, 270–76

Ng, L.W.C, D.P. Ng and W.P. Wong (2013), 'Is Supervised Exercise Training Safe in Patients with Anorexia Nervosa? A Meta-analysis', *Physiotherapy*, 99, 1–11

Ng, Marie, Tom Fleming, Margaret Robinson et al. (2014), 'Global, Regional, and National Prevalence of Overweight and Obesity in Children and Adults during 1980–2013: A Systematic Analysis for the Global Burden of Disease Study 2013', *Lancet*, May

Nicholls, Dasha, Deborah Christie, Louise Randall et al. (2001),
    'Selective Eating: Symptom, Disorder or Normal Variant', *Clinical
    Child Psychology and Psychiatry*, 6, 260–70
Nicholls, Dasha, Richard Lynn and Russell M. Viner (2011),
    'Childhood Eating Disorders: British National Surveillance Study',
    British Journal of Psychiatry, 198, 295–301
Nicholls, Dasha and Russell M. Viner (2009), 'Childhood Risk Factors
    for Lifetime Anorexia Nervosa by Age 30 Years in a National Birth
    Cohort', *Journal of the American Academy of Child and Adolescent
    Psychiatry*, 48, 791–9
Niklaus, Sophie, Vincent Boggio, Claire Chabanet et al. (2004),
    'A Prospective Study of Food Preferences', *Food Quality and
    Preference*, 15, 805–18
Nordin-Bates, Sanna M., Imogen Walker and Emma Redding (2011),
    'Correlates of Disordered Eating Attitudes Among Male and
    Female Young Talented Dancers: Findings From the UK Centres for
    Advanced Training', *Eating Disorders: The Journal of Treatment &
    Prevention*, 19:3, 211–33
Northstone, K., P. Emmett and the ALSPAC Study Team (2005),
    'Multivariate Analysis of Diet in Children at Four and Seven Years
    of Age and Associations with Socio-demographic Characteristics',
    European Journal of Clinical Nutrition, 59, 751–60

Onishi, Norimitsu (2008), 'Japan, Seeking Trim Waists, Measures
    Millions', *New York Times*, 13 June
Osman, Jamie L. and Jeffery Sobal (2006), 'Chocolate Cravings
    in American and Spanish Individuals: Biological and Cultural
    Influences', *Appetite*, 47, 290–301

Paltrow, Gwyneth (2013), *It's All Good: Delicious, Easy Recipes that
    will Make You Look and Feel Great*, London: Sphere
Pande, Rohini (2003), 'Selective Gender Differences in Childhood
    Nutrition and Immunization in Rural India: The Role of Siblings',
    Demography, 40, 395–418
Park, Min-Hae, Catherine Falconer, Helen Croker et al. (2014),
    'Predictors of Health-related Behaviour Change in Parents of
    Overweight Children in England', *Preventive Medicine*, 62,
    20–24

Patterson, Daniel (2013), *Coi: Stories and Recipes*, London: Phaidon Press

Paul, Candace, Keith E. Williams, Katherine Riegel et al. (2007), 'Combining Repeated Taste Exposure and Escape Prevention: An Intervention for the Treatment of Extreme Food Selectivity', *Appetite*, 49, 708–11

Peebles, Rebecka, Jenny L. Wilson and James D. Lock (2006), 'How do Children with Eating Disorders differ from Adolescents with Eating Disorders at Initial Evaluation?', *Journal of Adolescent Health*, 39, 800–805

Pember Reeves, Maud (1994), *Round About a Pound a Week*, London: Virago, facsimile of edn of 1913

Pitkeathley, Jill and David Emerson (1994), *Only Child: How to Survive Being One*, London: Souvenir Press

Pizzo, Bianca, Keith E. Williams, Candace Paul et al. (2009), 'Jump Start Exit Criterion: Exploring a New Model of Service Delivery for the Treatment of Childhood Feeding Problems', *Behavioral Intentions*, 24, 195–203

Planck, Nina (2007), *Real Food: What to Eat and Why*, London: Bloomsbury

Pliner, Patricia and Marcia Pelchat (1986), 'Similarities in Food Preferences between Children and their Siblings and Parents', *Appetite*, 7, 333–42

Pollan, Michael (2008), *In Defence of Food: The Myth of Nutrition and the Pleasures of Eating*, London: Allen Lane

Poncelet, Johan, Fanny Rinck and Fanny Bourgeat (2010), 'The Effect of Early Experience on Odor Perception in Humans: Psychological and Physiological Correlates', *Behavioural Brain Research*, 208, 458–65

Pooley, Siân (2009), 'Parenthood and Child-Rearing in England c. 1860–1910', Ph.D, University of Cambridge

Pooley, Siân (2010), 'All We Want is that Our Children's Health and Lives Should be Regarded: Child Health and Parental Concerns in England, c. 1860–1910', *Social History of Medicine*, 23, 528–48

Popkin, Barry (2006), 'Global Nutrition Dynamics: The World is Shifting Rapidly Toward a Diet Linked with Noncommunicable Diseases', *American Journal of Clinical Nutrition*, 84, 289–98

Popkin, Barry and Kiyah J. Duffey (2010), 'Does Hunger and Satiety Drive Eating Any More? Increasing Eating Occasions and Decreasing Time between Eating Occasions in the United States', *American Journal of Clinical Nutrition*, 91, 1342–7

Prentice, Andrew M. (2001), 'Fires of Life: The Struggles of an Ancient Metabolism in a Modern World', *Nutrition Bulletin*, 26, 13–27

Prescott, John (2012), *Taste Matters: Why We Like the Foods We Do*, London: Reaktion Books

Pritchard, Eric (1909), *The Physiological Feeding of Infants: A Practical Handbook of Infant Feeding*, London: Henry Kimpton

Puisais, J. and C. Pierre (1987), *Le Goût et l'Enfant*, Paris: Flammarion

Rapley, Gill (2008), *Baby-led Weaning: Helping your Baby to Love Good Food*, London: Vermilion

Remington, A., E. Añez, H. Croker et al. (2012), 'Increasing Food Acceptance in the Home Setting: A Randomized Controlled Trial of Parent-administered Taste Exposure with Incentives', *American Journal of Clinical Nutrition*, 95, 72–7

Resnicow, K. and S. Rollnick (2006), 'Motivational Interviewing for Pediatric Obesity: Conceptual Issues and Evidence Review', *Journal of the American Dietetic Association*, 106, 2024–33

Reverdy, C., F. Chesnel, P. Schlich et al. (2008), 'Effect of Sensory Education on Willingness to Taste Novel Food in Children', *Appetite*, 51, 156–65

Reverdy, C., P. Schlich, E.P. Köster et al. (2010), 'Effect of Sensory Education on Food Preferences in Children', *Food Quality and Preference*, 21, 794–804

Rhee, Kyung E., Julie C. Lumeng et al. (2006), 'Parenting Styles and Overweight Status in First Grade', *Pediatrics*, 117, 2047–55

Rice, Andrew (2010), 'The Peanut Solution', *New York Times*, 10 September

Roden, Claudia (1968), *A Book of Middle Eastern Food*, London: Penguin

Rodin, J., L.R. Silberstein and R. Striegel-Moore (1985), 'Women and Weight: A Normative Discontent', in T.B. Sonderegger (ed.), *Nebraska Symposium on Motivation: Vol. 32. Psychology and Gender*, 267–307, Lincoln: University of Nebraska Press

Rolls, Barbara (1986), 'Sensory-specific Satiety', *Nutrition Reviews*, 44, 93–101

Rolls, Barbara J., Elizabeth A. Bell and Bethany A. Waugh (2000a), 'Increasing the Volume of a Food by Incorporating Air Affects Satiety in Men', *American Journal of Clinical Nutrition*, 72, 361–8

Rolls, Barbara, Dianne Engell and Leann Birch (2000b), 'Serving Portion Size Influences 5 year old but not 3 year old Children's Food Intakes', *Journal of the American Dietetic Association*, 100, 232–4

Rolls, Barbara, Sion Kim-Harris and Marian W. Fischman (1994), 'Satiety after Preloads with Different Amounts of Fat and Carbohydrate: Implications for Obesity', *American Journal of Clinical Nutrition*, 60, 476–87

Rommel, Nathalie, Anne-Marie de Meyer, Louw Feenstra et al. (2003), 'The Complexity of Feeding Problems in 700 Infants and Young Children Presenting to a Tertiary Care Institution', *Journal of Pediatric Gastroenterology and Nutrition*, 37, 75–84

Rorty, Marcia, Joel Yager and Elizabeth Rossotto (2006), 'Why and How do Women Recover from Bulimia Nervosa? The Subjective Appraisals of Forty Women Recovered for a Year or More', *International Journal of Eating Disorders*, 14, 249–60

Roth, Michael P., Keith E. Williams and Candace M. Paul (2010), 'Treating Food and Liquid Refusal in an Adolescent with Asperger's Disorder', *Clinical Case Studies*, 9, 260–72

Rowan, Hannah and Cristen Harris (2012), 'Baby-led Weaning and the Family Diet. A Pilot Study', *Appetite*, 58, 1046–9

Rozin, Elizabeth (1994), *The Primal Cheeseburger*, New York: Penguin Books

Rozin, Paul (1969), 'Adaptive Food Sampling Patterns in Vitamin Deficient Rats', *Journal of Comparative and Physiological Psychology*, 69, 126–32

Rozin, Paul (1990), 'Acquisition of Stable Food Preferences', *Nutrition Reviews*, 48, 106–13

Rozin, Paul (1998a), *Towards a Psychology of Food Choice*, Brussels: Institut Danone

Rozin, Paul (2006), 'The Integration of Biological, Social, Cultural and Psychological Influences on Food Choices', in Richard Shepherd and Monique Raats, *The Psychology of Food Choice*, Wallingford: CABI

Rozin, P., R. Bauer and D. Catanese (2003), 'Attitudes to Food and Eating in American College Students in Six Different Regions of the United States', *Journal of Personality and Social Psychology*, 85, 132–41

Rozin, Paul, Sara Dow, Morris Moscovitch et al. (1998b), 'What Causes Humans to Begin and End a Meal? A Role for Memory for What has Been Eaten', *Psychological Science*, 9, 392–6

Rozin, P. and Deborah Schiller (1980), 'The Nature and Acquisition of a Chili Pepper Preference by Humans', *Motivation and Emotion*, 4, 77–101

Rozin, P. and T.A. Vollmecke (1986), 'Food Likes and Dislikes', *Annual Review of Nutrition*, 6, 433–56

Rundell, Maria (1827), *Domestic Economy and Cookery for Rich and Poor*, London: Longman, Rees, Orme, Brown and Green

Russell, Catherine Georgina and Anthony Worsley (2013), 'Why Don't They Like That? And Can I Do Anything about It? The Nature and Correlates of Parents' Attributions and Self-efficacy Beliefs about Preschool Children's Food Preferences', *Appetite*, 66, 34–43

Russell, Sharman Apt (2005), *Hunger: An Unnatural History*, New York: Basic Books

Salen, Arlene (1940), 'Hints on Infant Feeding', *The American Journal of Nursing*, 40, 649

Sandler, Lauren (2013), *One and Only: The Freedom of Having an Only Child and the Joy of Being One*, New York: Simon & Schuster

Savage, Jennifer S., Jennifer O. Fisher, Michele Marini et al. (2012), 'Serving Smaller Age-appropriate Entrée Portions to Children aged 3–5 yr Increases Fruit and Vegetable Intake and Reduces Energy Density and Energy Intake at Lunch', *American Journal of Clinical Nutrition*, 95, 335–41

Schaal, Benoist, Luc Marlier and Robert Soussignan (2000), 'Human Foetuses Learn Odours From Their Pregnant Mother's Diet', *Chemical Senses*, 25, issue 6, 729–37

Scheindlin, Benjamin (2005), '"Take One More Bite for Me": Clara Davis and the Feeding of Young Children', *Gastronomica*, 5, 65–9

Schnall, Simone (2007), 'Life as the Problem: Karl Duncker's Context', in Jaan Valsiner (ed.), *Thinking in Psychological Science: Ideas and Their Makers*, New Brunswick, New Jersey: Transaction Publishers

Schreck, Kimberly A., Keith Williams and Angela F. Smith (2004), 'A Comparison of Eating Behaviours between Children with and without Autism', *Journal of Autism and Developmental Disorders*, 34, 433–8

Seiverling, Laura, Amy Kokitus and Keith Williams (2012), 'A Clinical Demonstration of a Treatment Package for Food Selectivity', *Behavior Analyst Today*, 13, 1–6

Sela, Lee and Noam Sobel (2010), 'Human Olfaction: A Constant State of Change-blindness', *Experimental Brain Research*, 205, 13–29

Shephard, Sue (2001), 'A Slice of the Moon', in Harlan Walker (ed.), *Food and the Memory: Proceedings of the Oxford Symposium on Food and Cookery*, Totnes: Prospect Books

Shepherd, Gordon (2012), *Neurogastronomy: How the Brain Creates Flavor and Why It Matters*, New York: Columbia University Press

Shepherd, Richard and Monique Raats (2006), *The Psychology of Food Choice*, Wallingford: CABI

Singhal, Atul, I. Farooqi et al. (2002), 'Early Nutrition and Leptin Concentrations in Later Life', *American Journal of Clinical Nutrition*, 75, 993–9

Singhal, Atul, Kathy Kennedy et al. (2010), 'Nutrition in Infancy and Long-term Risk of Obesity: Evidence from 2 Randomized Control Trials', *American Journal of Clinical Nutrition*, 92, 1133–44

Sirikulchayanonta, C., P. Pavadhgul et al. (2010), 'Participatory Action Project in Reducing Childhood Obesity in Thai Primary Schools', *Asia Pacific Journal of Public Health*, 23, 917

Skinner, Jean D., Betty Carruth et al. (2002), 'Children's Food Preferences: A Longitudinal Analysis', *Journal of the American Dietetic Association*, 102, 11

Slater, Nigel (2004), *Toast: The Story of a Boy's Hunger*, London: Fourth Estate

Small, Dana M., Johannes C. Gerber, Erika Mak et al. (2005), 'Differential Neural Responses Evoked by Orthonasal versus Retronasal Odorant Perception in Humans', *Neuron*, 47, 593–605

Smink, Frédérique R.E., Daphne van Hoeken and Hans W. Hoek (2012), 'Epidemiology of Eating Disorders: Incidence, Prevalence', *Current Psychiatry Reports*, 14, 406–14

Smith, Lindsey, Katharine Conroy, Hongmai Wen et al. (2013), 'Portion Size Variably Affects Food Intake of 6-year old and 4-year old Children in Kunming, China', *Appetite*, 69, 31–8

Smith, Michelle I., Tanya Yatsunenko et al. (2013), 'Gut Microbiomes of Malawian Twin Pairs Discordant for Kwashiorkor', *Science*, 339, 548–54

Spahn, Joanne M., Rebecca S. Reeves and Kathryn S. Keim (2010), 'State of the Evidence Regarding Behavior Change Theories and Strategies in Nutrition Counseling to Facilitate Health and Food Behavior Change', *Journal of the American Dietetic Association*, 110, 879–91

Spieler, Marlena (2014), 'When a Food Writer Can't Taste', *New York Times*, 11 January

Spock, Benjamin (1946), *The Common Sense Book of Baby and Child Care*, New York: Duell, Sloan and Pearce

Steiner, Hans and James Lock (1998), 'Anorexia Nervosa and Bulimia Nervosa in Children and Adolescents: A Review of the Past 10 Years', *Journal of the American Academy of Child and Adolescent Psychiatry*, 37, 352–9

Steiner, J.E. (1979), 'Human Facial Expressions in Response to Taste and Smell Stimulation', *Advances in Child Development and Behavior*, 13, 257–95

Steinhausen, H.C. (2002), 'The Outcome of Anorexia Nervosa in the 20th Century', *American Journal of Psychiatry*, 159, 1284–93

Steinhausen, H.C. (2009), 'The Outcome of Bulimia Nervosa: Findings from One Quarter-century of Research', *American Journal of Psychiatry*, 166, 1331–41

Steinhausen, H.-Ch., C.R. Rauss-Mason and R. Seidel (1991), 'Follow-up Studies of Anorexia Nervosa: A Review of Four Decades of Outcome Research', *Psychological Medicine*, 21, 447–54

Stephen, Alison M. and Wald, Nicholas J. (1990), 'Trends in Individual Consumption of Dietary Fat in the United States, 1920–1984', *American Journal of Clinical Nutrition*, 52, 457–69

Stevens Bryant, Louise (1913), *School Feeding: Its History and Practice at Home and Abroad*, Philadelphia and London: J.B. Lippincott

Strauss, Stephen (2006), 'Clara M. Davis and the Wisdom of Letting Children Choose their own Diets', *Canadian Medical Association Journal*, 175, 1199–1201

Sullivan, S.A. and L.L. Birch (1990), 'Pass the Sugar, Pass the Salt: Experience Dictates Preference', *Developmental Psychology*, 26, 546–51

Sutton, David (2001), *Remembrance of Repasts: An Anthropology of Food and Memory*, Oxford: Berg

Sweetman, C., L. McGowan, H. Croker et al. (2011), 'Characteristics of Family Mealtimes Affecting Children's Vegetable Consumption and Liking', *Journal of the American Dietetic Association*, 111, 269–73

Tang, M.J.A. and A.J.A. Verboom (2014), 'Is Motivational Interviewing Effective as Treatment for Childhood Obesity?', *Appetite*, 76, 209

Tapper, Katy, Christine Shaw, Joanne Ilsley et al. (2009), 'Exploratory Randomised Control Trial of a Mindfulness-based Weight Loss Intervention for Women', *Appetite*, 52, 396–404

Tate, Deborah F., Robert W. Jeffery, Nancy E. Sherwood et al. (2007), 'Long-term Weight Losses Associated with Prescription of Higher Physical Activity Goals. Are Higher Levels of Physical Activity Protective against Weight Regain?', *American Journal of Clinical Nutrition*, 85, 954–9

Teicholz, Nina (2014), *The Big Fat Surprise: Why Butter, Meat and Cheese Belong in a Healthy Diet*, London and New York: Simon & Schuster

Tepper, Beverly J. (2008), 'Nutritional Implications of Genetic Taste Variation: The Role of PROP Sensitivity and Other Taste Phenotypes', *Annual Review of Nutrition*, 28, 367–88

That Sugar Film (2014), directed by Damon Gameau [film], USA: Madman Production Company

Thompson, Claire, Steven Cummins et al. (2014), 'What Does it Mean to be a "Picky Eater"? A Qualitative Study of Food-related Identities and Practices', *Appetite*, 84, 235–9

Thompson, Jan (2001), 'Prisoners of the Rising Sun: Food Memories of American POWs in the Far East During World War II', in Harlan Walker (ed.), *Food and the Memory: Proceedings of the Oxford Symposium on Food and Cookery*, Totnes: Prospect Books

Topham, Glade L., Laura Hubbs-Tait, Julie M. Rutledge et al. (2011), 'Parenting Styles, Parental Response to Child Emotion, and Family Emotional Responsiveness are Related to Child Emotional Eating', *Appetite*, 56, 261–4

Tovar, A., E. Hennessy, A. Pirie et al. (2012), 'Feeding Styles and Child
Weight Status among Recent Immigrant Mother–child Dyads',
International Journal of Behavioral Nutrition and Physical Activity,
9, 62–81

Ueland, Ø. (2007), 'Gender Differences in Food Choice', in Lynn
Frewer and Hans van Trijp (eds), *Understanding Consumers of
Food Products*, Abington: Woodhead Publishing

Ulander, Kerstin (2008), 'Healthier Eating Habits and Increased Food
Joy in the Elderly, Evaluation of the Development Effort, Diet,
Sensory, 70+', *Clinical Research*, 20, Kristianstad University, 1–32

Unusan, N. (2006), 'University Students' Food Preference and Practice
Now and during Childhood', *Food Quality and Preference*, 17,
362–8

Urbick, Bryan (2000), *About Kids: Foods and Beverages*, Leatherhead,
Surrey: Leatherhead Food Research

Urbick, Bryan (2011), 'Working with Children and Adolescents for
Food Product Development', in David Kilcast and Fiona Angus
(eds), *Developing Children's Food Products*, Cambridge: Woodhead

Valdes, J., F. Rodríguez-Artalejo, L. Aguilar et al. (2012), 'Frequency of
Family Meals and Childhood Overweight: A Systematic Review',
Pediatric Obesity, 8, E1–E13

Visser, Margaret (1991), *The Rituals of Dinner: the origins, evolution,
eccentricities and meaning of table manners*, London: Viking

Vollmer, Rachel L. and Amy R. Mobley (2013), 'Parenting Styles,
Feeding Styles, and their Influence on Child Obesogenic Behaviors
and Body Weight. A Review', *Appetite*, 71, 232–41

Walsh, Bryan (2013), 'Don't Blame Fat', *TIME*, 23 June, 29–35

Wansink, Brian (2004), 'Environmental Factors that Increase the Food
Intake and Consumption Volume of Unknowing Consumers',
Annual Review of Nutrition, 24, 455–79

Wansink, Brian (2011), *Mindless Eating: Why We Eat More Than
We Think We Do*, London: Hay House

Wansink, Brian, Matthew M. Cheney and Nina Chan (2003),
'Exploring Comfort Food Preferences across Age and Gender',
Physiology and Behavior, 79, 739–47

Wansink, Brian, J.E. Painter and J. North (2005), 'Bottomless Bowls: Why Visual Cues of Portion Size may Influence Intake', *Obesity Research*, 13, 93–100

Wardle, J., L.J. Cooke, E.L. Gibson et al. (2003a), 'Increasing Children's Acceptance of Vegetables; A Randomized Trial of Parent-led Exposure', *Appetite*, 40, 155–62

Wardle, J. and L. Cooke (2008), 'Genetic and Environmental Determinants of Children's Food Preferences', *British Journal of Nutrition*, 99, S15–S21

Wardle, Jane and Lucy J. Cooke (2010), 'One Man's Meat is Another Man's Poison', *European Molecular Biology Association*, EMBO reports, 1–6

Wardle, J., M.L. Herrera, L.J. Cooke et al. (2003b), 'Modifying Children's Food Preferences: The Effects of Exposure and Reward on Acceptance of an Unfamiliar Vegetable', *European Journal of Clinical Nutrition*, 57, 341–8

Washington, Booker T. (2008), *Up from Slavery: An Autobiography*, Oxford: Oxford University Press

Webb, Thomas L., Paschal Sheenan and Christopher Armitage (2006), 'Implementation Intentions: Strategic Automization of Food Choice', in Richard Shepherd and Monique Raats (eds), *The Psychology of Food Choice*, Wallingford: CABI

Weber, Eugen (1981), 'Fairies and Hard Facts: The Reality of Folktales', *Journal of the History of Ideas*, vol. 42, no. 1, 93–113

Weston, Janet A. and Mark Colloton (1993), 'A Legacy of Violence in Nonorganic Failure to Thrive', *Child Abuse and Neglect*, 17, 709–14

Wildes, Jennifer E., Nancy L. Zucker and Marsha D. Marcus (2012), 'Picky Eating in Adults: Results of a Web-Based Survey', *International Journal of Eating Disorders*, 45, 575–82

Wilkinson, Michelle Lynn, Austin Lane Brown, Walker Seward Poston et al. (2014), 'Physician Weight Recommendations for Overweight and Obese Firefighters, United States, 2011–2012', *Preventing Chronic Disease*, 11, 140091

Williams, Keith (2011), 'Increasing Children's Food Choices: Strategies Based upon Research and Practice', in David Kilcast and Fiona Angus (eds), *Developing Children's Food Products*, Cambridge: Woodhead

Williams, Keith, Candace Paul, Bianca Pizzo et al. (2008), 'Practice Does Make Perfect. A Longitudinal Look at Taste Exposure', *Appetite*, 51, 739–42

Wilson, Bee (2002), 'Dairylea Lunchables', *New Statesman*, 2 December

Wilson, Bee (2005), 'The Weight of the World', *Sunday Telegraph*, 4 December

Wilson, Bee (2009), *Swindled: The Dark History of Food Fraud*, Princeton: Princeton University Press

Wilson, Bee (2012), *Consider the Fork: A History of How We Cook and Eat*, New York: Basic Books

Wing, Rena R. and Suzanne Phelan (2005), 'Long-term Weight Loss Maintenance', *American Journal of Clinical Nutrition*, 82, 222S–5S

Wise, Roy A. (2001), 'The Neurobiology of Food Craving', in Marion Hetherington (ed.), *Food Cravings and Addiction*, Leatherhead: Leatherhead Food Publishing

Wise, Roy A. (2006), 'Role of Brain Dopamine in Food Reward and Reinforcement', *Philosophical Transactions of the Royal Society*, 361, 1149–58

Wright, C.M., K. Cameron, M. Tsiaka et al. (2011), 'Is Baby-Led Weaning Feasible? When do Babies First Reach out for and Eat Finger Foods?', *Maternal and Child Nutrition*, 7, 27–33

Xia, Wei, Xin Zhang, and Jiajia Wang (2012), 'Survey of Anaemia and Helicobacter Pylori Infection in Adolescent Girls in Suihua, China and Enhancement of Iron Intervention Effects by H. Pylori Eradication', *British Journal of Nutrition*, 108, 357–62

Yeomans, Martin and Lucy Chambers (2011), 'Satiety-relevant Sensory Qualities Enhance the Satiating Effects of Mixed Carbohydrate-Protein Preloads', *American Journal of Clinical Nutrition*, 94, 1410–17

Zajonc, Robert B. (1968), 'Attitudinal Effects of Mere Exposure', *Journal of Personality and Social Psychology*, 9, 1–27

Zajonc, Robert B. (1980), 'Feeling and Thinking: Preferences Need no Inferences', *American Psychologist*, 35, 151–75

Zajonc, Robert B. and Hazel Markus (1982), 'Affective and Cognitive Factors in Preferences', *Journal of Consumer Research*, 9, 123–31

Zeinstra, Gertrude, M.A. Koelen, F.J. Kok et al. (2009), 'Children's Hard-wired Aversion to Pure Vegetable Tastes. A "failed" Flavour–Nutrient Learning Study', *Appetite*, 52, 528–30

Zhang, Gen-Hua, Meng-Ling Chen, Si-Si Liu et al. (2011), 'Effects of Mother's Dietary Exposure to Acesulfame-K in Pregnancy or Lactation on the Adult Offspring's Sweet Preference', *Chemical Senses*, 36, 763–70

Zocca, Jaclyn M., Lauren B. Shomaker et al. (2011), 'Links between Mothers' and Children's Disinhibited Eating and Children's Adiposity', *Appetite*, 56, 324–31

Zucker, Nancy L., Kevin S. LaBar et al. (2007), 'Anorexia Nervosa and Autism Spectrum Disorders: Guided Investigation of Social Cognitive Endophenotypes', *Psychological Bulletin*, 133, 976–1006

Notes

Introduction

1 Jelliffe (1962).
2 De Sa et al. (2013).
3 Cornwell and McAlister (2010).
4 Lim et al. (2012).
5 Moss (2014).
6 Hoek and Hoeken (2003).
7 Rozin, Bauer and Catanese (2003).
8 Lustig et al. (2012), Lustig (2014), Pollan (2008), Walsh (2013).
9 Teicholz (2014).
10 Nestle et al. (1998), p. S51, Morbidity and Mortality Weekly Report (1994) and Stephen and Wald (1990).
11 Katz and Meller (2014).
12 Katz (2014).
13 Walsh (2014).
14 Köster and Mojet (2007).
15 Pollan (2008).
16 Garcia et al. (2009).
17 Wilkinson et al. (2014).
18 Hare (2010).
19 Wise (2006).
20 Drewnowksi et al. (2012).
21 Lustig et al. (2012).
22 Leigh Gibson (2001).
23 Leigh Gibson (2001), Wise (2006).
24 Wise (2006).
25 Leigh Gibson (2001).
26 Cornwell and McAlister (2011).
27 Unusan (2006).
28 http://www.hopkinsmedicine. org/healthlibrary/conditions/ digestive_disorders/ constipation_85,P00363/, accessed November 2014.
29 Rozin and Schiller (1980).
30 Baumeister et al. (1998).
31 Köster, Rummel et al. (2001).
32 Köster (2009).
33 Gameau (2014).
34 Meiselman (2006), pp. 183–4.

CHAPTER 1: Likes and Dislikes

1 Rozin and Vollmecke (1986), p. 435.
2 See Havermans (2011), Berridge (2009) and Castro and Berridge (2014).
3 Berridge (2009).
4 See, for example, Havermans (2011), Wise (2006).

5 Berridge (2009).
6 Wise (2006).
7 Catanzano et al. (2013), Kaminski et al. (2000), Tepper (2008).
8 Llewellyn et al. (2010).
9 Hales and Barker (2001).
10 Strauss (2006).
11 Davis (1939).
12 Davis (1939).
13 Davis (1939).
14 Davis (1928).
15 Strauss (2006).
16 Scheindlin (2005); Stephen Strauss conducted an interview with Donald's widow in 2001 that confirmed that he was always a 'good eater' (email from Strauss to the author, July 2014).
17 See, for example, Goldberg (1990), Planck (2007), Spock (1946); see Birch (1999) for a refutation of the conclusion that Davis's work supports the 'wisdom of the body'.
18 Davis (1939).
19 Scheindlin (2005); see also Bentley (2006), pp. 72–4.
20 Spock (1946).
21 Hirschmann and Zaphiropoulos (1985).
22 http://www.babyledweaning.com/features/random-stuff/a-modern-take-on-the-clara-m-davis-paper/, accessed November 2014.
23 Leigh Gibson (2001).
24 Rozin (1969).
25 Leigh Gibson (2001), p. 203.
26 Faith et al. (2006) and Faith et al. (2012).
27 Prescott (2012) pp. 175–6.
28 Fildes et al. (2014).
29 Breen, Plomin and Wardle (2006).
30 Fallon, Rozin and Pliner (1984).
31 Wardle and Cook (2010).
32 Sullivan and Birch (1990).
33 Rozin and Vollmecke (1986), p. 437.
34 Rozin (2006).
35 'Sweet Sensations', Guardian, Friday, 17 January 2014.
36 Tepper (2008).
37 Bartoshuk (2000), Dinehart et al. (2006), Duffy et al. (2004).
38 Dinehart et al. (2006).
39 See for example Anliker et al. (1991).
40 http://www.jancisrobinson.com/articles/the-prop-test-and-reactions-to-it, accessed December 2014.
41 Discussed in Tepper (2008).
42 Feeney et al. (2014).
43 Catanzaro et al. (2013).
44 Catanzaro et al. (2013).
45 Burd et al. (2013).
46 Conversation with the author, October 2013.
47 Zajonc (1968).
48 Zajonc (1980); see also Zajonc and Markus (1982) for a discussion of how 'mere exposure' plays out in food preferences.
49 Birch and Marlin (1982).
50 Skinner et al.(2002).
51 Prescott (2012).
52 Prescott (2012).
53 Russell and Worsley (2013).
54 Añez et al. (2012), Carnell et al. (2011), Cooke et al. (2011), Wardle et al. (2003a and

2003b), Wardle and Cooke (2008) and (2010).
55 Harris (2008).
56 Maier, Chabanet et al. (2007).
57 Coulthard, Harris et al. (2014).
58 http://www.unicef.org.uk/BabyFriendly/About-Baby-Friendly/Breastfeeding-in-the-UK/UK-Breastfeeding-rates/, accessed March 2015.
59 http://www.cdc.gov/breast feeding/pdf/2014breastfeeding reportcard.pdf, accessed March 2015.
60 Karmel (1991).
61 http://www.weightconcern.org.uk/tinytastes, accessed November 2014.
62 Ernsperger and Stegen-Hanson (2004).
63 Schreck et al. (2004), Ernsperger and Stegen-Hanson (2004).
64 Paul et al. (2007).
65 Duncker (1941).
66 Duncker (1938).
67 Cruwys et al. (2015).
68 Schnall (2007).
69 Duncker (1938).
70 Duncker (1938).
71 Schnall (2007).
72 Zeinstra et al. (2009).
73 Mojet and Köster (2006), Lévy, MacRae and Köster (2006).

CHAPTER 2: **Memory**

1 Spieler (2014).
2 Small et al. (2005).
3 http://www.anosmia foundation.com/suffer.shtml, accessed March 2015.
4 Conversation with the author, January 2014.
5 http://www.bbc.co.uk/programmes/b01r95hj, accessed November 2014.
6 Rozin et al. (1998b).
7 Rozin et al. (1998b).
8 Leigh Gibson (2001).
9 Wise (2006).
10 Unusan (2006).
11 Steiner (1979).
12 Schaal et al. (2000).
13 Mennella, Johnson and Beauchamp (1995).
14 Zhang et al. (2011).
15 Gugusheff et al. (2013).
16 Mennella and Beauchamp (1991), (1993), Mennella et al. (1995) and (2005).
17 'Bad Eating Habits Start in the Womb', *New York Times*, 1 December 2013.
18 Talk given by Dr Lucy Cooke at NH Live conference, 2013 'Understanding Young Children's Food Preferences'.
19 Beauchamp and Mennella (2011).
20 Salen (1940).
21 Baby Centre, 'Vanilla Natural Flavoring in babies bottles', http://community.babycentre.co.uk/post/a23870045/vanilla_natural_flavoring_in_babies_bottles, accessed June 2015.
22 Shen et al. (2014).
23 http://abcnews.go.com/Health/Diabetes/mead-johnson-drops-chocolate-flavored-emfagrow-parent-uproar/story?id=10876301, accessed November 2014.
24 Haller et al. (1999).
25 Lobstein (1988).
26 Sela and Sobel (2010).

27 Levin Pelchat and Blank (2001).
28 Malnic et al. (1999).
29 Shepherd (2012).
30 Shepherd (2012).
31 Shephard (2001).
32 Thompson (2001).
33 Prescott (2012).
34 Sutton (2001).
35 Sutton (2001).
36 Carafoli (2001).
37 Mendelson (2013).
38 Blumenthal (2009).
39 Blake (2001), Dalton et al. (2000).
40 Patterson (2013).
41 Patterson, conversation with the author, February 2014.
42 Sutton (2001).
43 Bittman (2013).
44 Dutton (1906).
45 Global Ice Cream, October 2014, MarketLine Industry Profile.

CHAPTER 3: Children's Food

1 Quoted in Elliott (2008).
2 Quoted in Clifton and Spencer (1993).
3 Hecht (1912) and Hecht (1913).
4 McMillan (1909).
5 Stevens Bryant (1913).
6 Hecht (1912).
7 Hecht (1912), p. 89.
8 McMillan (1909).
9 Crowley (1909).
10 Hecht (1913).
11 *Financial Times*, 6 December 2013.
12 *Evening News*, 14 May 1912.
13 Culpeper (1662).
14 Visser (1991), p. 46.
15 Washington (2008).
16 Crowley (1909).
17 Stevens Bryant (1913).
18 Stevens Bryant (1913).
19 Pember Reeves (1994).
20 Pember Reeves (1994).
21 Pember Reeves (1994).
22 Hecht (1913), p. 310.
23 Pooley (2009) and (2010).
24 Albala (2002).
25 Dutton (1906), p. 15.
26 Dutton (1906), p. 17.
27 Dutton (1906), p. 23.
28 Clark (1874).
29 Holt (1923).
30 Holt (1923).
31 Rundell (1827).
32 Pooley (2009) and (2010).
33 Pritchard (1909).
34 Hecht (1912), pp. 304–5.
35 David (2000).
36 Clifton and Spencer (1993).
37 Quoted in Hardyment (1995), p. 264.
38 Boorstin (2001).
39 Boorstin (2001).
40 Groves (2002).
41 Kawash (2013).
42 Cathro and Hilliam (1994).
43 Castonguay et al. (2013).
44 Groves (2002), p. 119.
45 Elliott (2008).
46 Hilliam (1996).
47 Urbick (2000), p. 65.
48 Wilson (2002).
49 Urbick (2011), p. 219.
50 Urbick (2011), p. 219.
51 Urbick (2000), p. 11.
52 Lobstein (1988), p. 40.
53 Lobstein (1988), p. 48.
54 Williams (2011), p. 135.
55 Jennings (2009).

56 The Associated Press, 28 August 2013, 'Some schools drop out of new healthy federal lunch program, citing small portions and foods kids won't eat' http://www.nydailynews.com/life-style/health/schools-drop-new-healthy-federal-lunch-program-article-1.1439576, accessed June 2015.

57 http://www.theguardian.com/lifeandstyle/2009/jan/30/family1, accessed March 2015.

58 See Wansink (2002).

59 Mead (1943).

60 David (2000).

61 Popkin (2006).

62 Skinner (2002).

63 Humble (2010).

CHAPTER 4: Feeding

1 Cole and Lanham (2011).

2 Block and Krebs (2005).

3 Weston and Colloton (1993).

4 Goh (2009).

5 Jiang Jingxiong et al. (2007).

6 Jiang Jingxiong et al. (2007).

7 Hecht (1912), pp. 33–4.

8 Prentice (2001).

9 Prentice (2001).

10 Jiang Jingxiong et al. (2007).

11 Baldeesh Rai, 'Asian Diets and Cardiovascular Disease', paper given to 2013 Nutrition and Health Live conference, London.

12 Ng et al. (2014).

13 French and Crabbe (2010).

14 Pollan (2008).

15 Jiang Jingxiong et al. (2007).

16 Beecher (1986).

17 Beecher (1986).

18 Bentley, Wasser and Creed-Kanashiro (2011).

19 Birch (1998), Birch (1999), Birch and Anzman (2010).

20 Batsell et al. (2002).

21 Holt (1923).

22 Hubble and Blake (1944), p. 447.

23 Clifton and Spencer (1993).

24 Batsell et al. (2002).

25 Carnell, Cooke et al. (2011).

26 Galloway et al. (2006).

27 Galloway et al. (2006).

28 Discussed in Vollmer and Mobley (2013).

29 Vollmer and Mobley (2013).

30 Tovar et al. (2012).

31 Vollmer and Mobley (2013).

32 Rhee et al. (2006).

33 Vollmer and Mobley (2013).

34 Hoerr et al. (2009).

35 Huang, Parks et al. (2012).

36 Carnell, Cooke et al. (2011).

37 Fisher and Birch (2002).

38 Vollmer and Mobley (2013).

39 Topham et al. (2011).

40 www.ellynsatterinstitute.org, accessed December 2014.

41 Rapley and Murkett (2008) and www.rapleyweaning.com, accessed December 2014.

42 Rapley and Murkett (2008).

43 http://www.rapleyweaning.com/assets/blw_guidelines.pdf, accessed March 2015.

44 Rowan and Harris (2012).

45 Wright et al. (2011).

46 Gold (1993).

47 http://www.schoolfoodplan.com/wp-content/uploads/2013/07/School_Food_Plan_2013.pdf, accessed March 2015.

48 Farris et al. (2014).

49 http://www.childrensfood
    trust.org.uk/news-and-events/
    news/school-meals-help-fussy-
    children-try-new-foods,
    accessed December 2014.
50 Itoh (2011).

CHAPTER 5: **Brothers and Sisters**

1   Levin and Kirby (2012).
2   Pliner and Pelchat (1986).
3   De Leeuw et al. (2007).
4   Smith, Yatsunenko et al.
    (2013); see also 'Debugging
    the Problem', *Economist*,
    2 February 2013.
5   http://timesofindia.indiatimes.
    com/india/India-deadliest-
    place-in-world-for-girl-child/
    articleshow/11707102.cms,
    accessed December 2014.
6   Pande (2003).
7   Pande (2003).
8   Pande (2003).
9   Pande (2003).
10 Weber (1981).
11 Weber (1981).
12 Fong (2004).
13 Sandler (2013).
14 Laybourn (1994).
15 Sandler (2013).
16 Mentioned in Coates (1996).
17 Quoted in Pitkeathley and
    Emerson (1994).
18 Bourdieu (1986).
19 Brillat-Savarin (2009).
20 Bourdieu (1986).
21 Cathro and Hilliam (1994).
22 Conley and Glauber (2007).
23 Blisset et al. (2006); but see
    also Hendy and Williams
    (2012) for a paper suggesting
    that parents do not always

feed children of different sexes
differently.
24 Bauer et al. (2011).
25 Hammons and Fiese (2011),
    Valdes et al. (2012).
26 Armstrong and Janicke (2012).
27 Neumark-Sztainer et al. (2010).
28 Bauer et al. (2011).
29 Slater (2004).
30 'Holding Back Half the Nation',
    Economist, 29 March 2014.
31 http://www.nhs.uk/chq/
    Pages/how-many-calories-do-
    teenagers-need.aspx?Category
    ID=51&SubCategoryID=165,
    accessed September 2014.
32 Köster (2003).
33 Urbick (2011).
34 Discussed in Ueland (2007).
35 Wansink et al. (2003).
36 See Kimura (2009) and (2012)
    for gender stereotypes of food
    in Japan.
37 Komatsu (2008).
38 Martens (1997).
39 Eftekhari et al. (2009).
40 Hercberg et al. (2001).
41 Xia et al. (2012).
42 http://healthyeating.sfgate.
    com/should-eat-liver-iron-
    intake-3367.html, accessed
    September 2014.
43 Nelson (1996), p. 362.
44 Eftekhari et al. (2009).
45 Eftekhari et al. (2009).
46 Dr Laura Stewart, 'An Update
    on Obesity in the U.K. Young'
    Nutrition and Health Live,
    London, http://www.nutrition
    andhealth.co.uk/.
47 Jain et al. (2001).
48 Kuchler and Variyam (2003).
49 Howard et al. (2008).

50 Rozin et al. (2003).
51 Geier and Rozin (2008).
52 Rodin et al. (1985).
53 Ueland (2007).
54 Cited in Groves (2002).
55 Sirikulchayanonta et al. (2010).
56 Musaiger et al. (2012).
57 http://www.arabtimesonline.
   com/Default.aspx?TabId=96
   &smid=414&ArticleID=162
   009&reftab=36&t=Kuwait-
   lifestyle-could-lead-to-obesity,
   accessed March 2015.
58 Roden (1968).
59 Musaiger at al. (2013).
60 Botz-Bornstein and Abdullah-
   Khan (2014).
61 Musaiger et al. (2013).
62 Urbick (2011).
63 Hormes and Rozin (2009).
64 Osman and Sobal (2006).
65 Kuijer and Boyce (2014).

CHAPTER 6: Hunger

1 https://www.nokidhungry.org/
   solution/ending-childhood-
   hunger, accessed December
   2014.
2 http://www.feedingamerica.org/
   hunger-in-america/impact-of-
   hunger/hunger-and-poverty/,
   accessed December 2014.
3 Ficker and Graves (1971),
   p. 44.
4 http://www.wfp.org/hunger/
   stats, accessed December 2014.
5 Cutts et al. (2011).
6 Stevens Bryant (1913), p. 219.
7 Carlson (1993), p. 6.
8 Mattes (1990) and (2010).
9 Mattes (2010).
10 Mattes (2010).
11 De Graaf et al. (2004).
12 De Graaf et al. (2004).
13 Kovacs et al. (2002).
14 Kissileff et al. (2003).
15 De Graaf et al. (2004).
16 Benelam (2009).
17 Benelam (2009).
18 De Graaf et al. (2004).
19 Carlson (1993).
20 Keys, Brožek et al. (1950).
21 Brožek (1953).
22 Hoefling et al. (2009).
23 Rice (2010).
24 http://futurefood2050.com/
   peanut-butter-that-saves-lives/,
   accessed December 2014.
25 http://www.unicef.org/
   bangladesh/Child_and_Mother
   _Nutrition_Survey.pdf,
   accessed April 2015.
26 Conversation with the author,
   March 2014.
27 Ali et al. (2013).
28 Benelam (2009).
29 Cathro and Hilliam (1994).
30 Paltrow (2013).
31 See, for example, Yeomans and
   Chambers (2011).
32 Benelam (2009).
33 Rolls et al. (2000a).
34 Rolls et al. (2000a).
35 Mattes (2005).
36 Prescott (2012).
37 Mattes (2005).
38 Popkin and Duffy (2010).
39 Lehmann (2003).
40 Evers et al. (2013).
41 Rolls et al. (2000b).
42 Savage, Fisher et al. (2012).
43 Smith, Conroy et al. (2013).
44 Smith, Conroy et al. (2013).
45 Wansink et al. (2005).
46 Nestle (2007).

47 Wansink (2011).
48 Wansink (2011).
49 Discussed in Benelam (2009).
50 Rolls (1986).
51 Johnson (2000).
52 Johnson (2000).
53 Johnson (2000).
54 Tapper (2009).
55 Alberts et al. (2010).

CHAPTER 7: **Disorder**

1 Thompson et al. (2014).
2 Correspondence between Claire Thompson and the author, November 2014.
3 Thompson et al. (2014).
4 Rozin et al. (2003).
5 Zucker et al. (2007).
6 Herzog et al. (1999).
7 Delaney et al. (2014).
8 Bryant-Waugh et al. (2010).
9 Rommel et al. (2003).
10 Kauer et al. (2015).
11 Conversation with the author, May 2014.
12 Nicholls et al. (2001).
13 Conversation with the author, May 2014.
14 Bryant-Waugh (2013).
15 Murray et al. (2013).
16 Seiverling et al. (2012).
17 Seiverling et al. (2012).
18 Seiverling et al. (2012).
19 Roth et al. (2010).
20 Arnold (2012).
21 Baron-Cohen et al. (2013).
22 Zucker et al. (2007).
23 Baron-Cohen et al. (2013).
24 Discussed in Hay and Sachdev (2011).
25 Arnold (2012).
26 Nicholls et al. (2011).
27 Marshall (1895).
28 Nordin-Bates et al. (2011).
29 Ng et al. (2013).
30 Klump (2013).
31 Conversation with the author, May 2014.
32 Steinhausen (1991).
33 Quoted in Bryant-Waugh and Lask (2013).
34 Steinhausen (2002).
35 Lock and Le Grange (2004).
36 Bruch (1978).
37 Lock and Le Grange (2004).
38 Brown (2009).
39 Wilson (2005).
40 Steinhausen (2009).
41 Rorty et al. (2006).
42 Bailer et al. (2004).
43 Moore (2011).
44 Moore (2011).
45 Gopnik (2011).
46 Zucker et al. (2007).

CHAPTER 8: **Change**

1 Kushner (2012).
2 Ng et al. (2014).
3 Onishi (2008).
4 Kushner (2012).
5 Kushner (2012).
6 'Slurp! Revealing the History of Ramen', Talk by Barak Kushner to the Guild of Food Writers, London, 18 July 2013.
7 Collingham (2011).
8 Kushner (2012).
9 Collingham (2011).
10 Cwiertka (2006).
11 Collingham (2011).
12 Ishige (2001).
13 Kushner (2012).
14 Ishige (2001).
15 Rozin (1994).

16 Henry (2014).
17 Miller and Rollnick (2013).
18 Miller and Rollnick (2013).
19 Spahn et al. (2010); see also Resnicow and Rollnick (2006).
20 Bowen et al. (2002).
21 Tang and Verboom (2014).
22 Miller and Rollnick (2013).
23 Chapman and Ogden (2010).
24 http://www.thetimes.co.uk/tto/health/news/article4425583.ece, accessed April 2015.
25 Chapman and Ogden (2010).
26 Mentioned in Webb et al. (2006).
27 Appelhans et al. (2014).
28 Lucas et al. (2013).
29 Comments made by Baldeesh Rai after her presentation 'Asian Diets and Cardiovascular Disease' at NHLive, London, November 2013.
30 Wing and Phelan (2005), Elfhag and Rössner (2005).
31 Anderson et al. (2007b).
32 Anderson et al. (2007a).
33 Elfhag and Rössner (2005).
34 Kayman et al. (1990).
35 Kayman (1990).
36 Drewnowski (1997).
37 Shepherd (2012), Gonzalez et al. (2008).
38 Mattes (1997).
39 Mattes (1997).
40 Itard (1932).
41 Puisais and Pierre (1987).
42 Reverdy et al. (2010).
43 Koistinen and Ruhanen (2009).
44 http://www.peda.net/veraja/projekti/saperemenetelma, accessed December 2014.
45 Koistinen and Ruhanen (2009).
46 Email from Arja Lyytikäinen to the author, April 2014.
47 Koistinen and Ruhanen (2009).
48 Reverdy et al. (2008) and (2010), Mustonen and Tuorila (2010).
49 Reverdy et al. (2010).
50 Mustonen and Tuorila (2010).
51 Keller et al. (2005).
52 Hughes et al. (2004).
53 Ulander (2008).
54 Email to the author from Albert Westergren, February 2015.
55 Rozin and Schiller (1980).

Acknowledgements

This book has benefited greatly from the knowledge and wisdom of many others.

I'm so grateful to those who spoke to me and shared either their experiences or their scholarship on the questions of why and how we come to eat as we do. Some spoke anonymously. Of those who spoke or emailed on the record, I'd like to thank Jose Luis Álvarez Morán and all at Action Against Hunger, Duncan Boak and all at Fifth Sense, Paul Breslin, Lucy Cooke, Helen Crawley, Luis Gigliotti, Yasmin Hosny, David Jukes, E.P. Köster, Barak Kushner, Arja Lyytikäinen, A.O. Musaiger, Daniel Patterson, Dympna Pearson, Dawn and Abi Millard, Marlena Spieler, Carmel McConnell and all at Magic Breakfast, Susan Ringwood, Stephen Strauss, Claire Thompson, Albert Westergren, Keith Williams.

My thinking on food always owes much to my friends at the Oxford Symposium on Food and Cookery. I presented an early version of the ideas in *First Bite* to The School of Artisan Cookery; thank you to Alison Swan Parente for inviting me to talk. Another institution I need to thank is the Guild of Food Writers for, among other things, putting on Barak Kushner's excellent ramen noodle workshop.

I'm always grateful to the staff of the Cambridge University Library, where I did much of the research; and to the staff of Hot Numbers coffee shop, where I did much of the writing.

I've also been helped by conversations with various people. I found that this was a subject that everyone has a view on. First and foremost, my thanks go to David, Tasha, Leo and Tom Runciman (thanks, Tom, for being so brutally honest). Others who gave me insights and other help include, in no particular order, Abby Scott, Lily Scott Turner, Mark Turner, Melissa Mohr and Caspar Hare, Ranjita Lohan and family, Sarah and Olivia Ray, Dan Jones and Sophie Hannah, Helen Conford, Diana Henry, Jane Kramer, Caroline Boileau, Catherine Blyth, Deborah Friedell, Anne Malcolm, Imogen Roth, Freya and Psyche Brackett, Amy Bryant, Ed Caffyn, Hilary Cooper, Sybil Kapoor, Emily Gowers, Michele Humes, Susan Friedland, Gareth Stedman Jones, Rose Hilder, Anna Hont, Attila Bacsò, Cara Isaac, Lizzie Collingham, Sharon Knights, Jane Ladlow, Anthea Morrison, Tamsin O'Connell, Siân Pooley, Ruth and Garry Runciman, Cathy Runciman, Gonzalo Gil, Lisa Runciman and Reg Lee, Ruth Scurr, Catherine Carr, Inigo Thomas, Andrew Wilson and Katherine Duncan-Jones. Special thanks to Emily Wilson, Caroline Boileau and Miranda Landgraf for reading draft versions and offering perceptive comments. I'm grateful to Sylvana Tomaselli for telling me that my original idea didn't work; you were right.

Thank you to the editors of my former food column at *Stella* magazine, especially Elfreda Pownall who guided my thoughts on food over many years.

I owe huge thanks to my agents, Zoe Pagnamenta and Sarah Ballard plus Zoe Ross, who couldn't have been more supportive, of me and of this project.

It's been a joy to work with the illustrator Annabel Lee on a second book; she drew the birthday cake of my childhood dreams. Thank you so much to designer Jo Walker for the brilliant cover.

I've been very fortunate to work for a second time with Lara Heimert at Basic Books; and to work for the first time with Louise Haines at Fourth Estate: two wonderful editors who kept me on my mettle. They each improved the book in so many ways, large and small. Kathy Streckfus in the US and Morag Lyall in the UK were both superb copyeditors. I'd also like to thank all at Basic Books including Michelle Welsh-Horst, Leah Stecher, Melissa Raymond, Cassie Nelson and all at Fourth Estate including Georgia Mason, Jo Walker, and Patrick Hargadon. Needless to say, the mistakes are my own.

Index

Abraham G 35, 36
Action Against Hunger 241
Africa 240
Allen, John S. 304
almond oil 2
Álvarez Morán, Jose Luis 240–1, 242, 243
anaemia 212–13
Anderson, Dr James 329
anorexia
  age of sufferers 286–8
  and being hungry 6, 231
  causes of 286–90
  and cognitive impairment 285–6
  as disconnected from internal experiences 300–1
  focus on symptoms 293
  learning how to eat 300
  and losing weight 265
  Maudsley approach (or family based treatment) 293–7
  mortality rates 266, 293
  prevailing view 284–5
  and puberty 290–1
  recovery from 267, 291–3
  refeeding process in later life 297–8
  see also bulimia; eating disorders
anosmia 73–8

Applebee's 220
Arnold, Carrie 286
Ashputtel 197
Asia 195, 210
Asperger's syndrome 281–2
Athena 97
Auschwitz 93
Australia 217
Austria 299
autism 61–4, 271, 285
Axel, Richard 88

babies
  and breast milk 2, 39, 57–8, 83–5, 119
  development of taste buds 81–2
  and flavour window 57–9
  food given to 139–40, 142
  and formula milk 2, 58, 84–7, 154
  reactions to basic tastes 7, 81
  in utero memories 82–3
Baby and Child Care (Dr Spock) 38, 336
Baby Led Weaning (BLW) 183–6
Ballet Shoes (Streatfeild) 191
Bangladesh 240–1, 242, 328
Barker, David 31, 163
Baron-Cohen, Simon 284–5

Bartoshuk, Linda 45
Bauer, Katherine 206
Baumeister, Roy 20–1
Beauchamp, Gary 83, 84, 100
beetroot 4, 70–1
bento lunches 188
benzaldehyde 101
Berridge, Kent 29–30
Beth (adult anorexic) 298
binge eating 29, 330
Birch, Leann 170, 175, 180–1
birthday cake 148–9, 155
The Bitter Cry of Children (Spargo) 121
bitter foods 44–50, 58, 144–5
Bittman, Mark 106
Bloomberg, Mayor 307
Blumenthal, Heston 100
Blyton, Enid 191
BMI (Body Mass Index) 48, 49, 164, 179, 216, 258, 324
Boak, Duncan 77
Boston 162–3
Bourdieu, Pierre 203, 204
Braddon, Russell 95
Bradford 114–17, 121
Brave Girl Eating (Brown) 296
Bread and Jam for Frances (Hoban) 51
breakfast cereal *see* cereal
Breslin, Dr Paul 100–1
Briend, André 239
Brillat-Savarin, Jean Anthelme 204
Brown, Harriet 296
Brown, Kitty 296–7
Brown, Margaret Wise 129
Bruch, Hilde 293–4
Bryant, Louise Stevens 122
Buck, Linda 88
bulimia 6, 205, 265, 267, 298–300
　see also anorexia; eating disorders
Buna 93

Cake (Humble) 148
calories 6, 8, 9, 209, 214, 242, 249, 251
Canada 344
Candy: A Century of Panic and Pleasure (Kawash) 136
Cantines Scolaires 112
Carafoli, John F. 98
Carlson, Anton J. 231
Carolina Population Center 250
CCK 234
Central America 210
cereal 2, 129, 136, 137, 165, 213, 221, 246, 260–1
Chaga tribe (Tanzania) 120
Chang, David 144
changing diets 22–3
　and culture 328
　and exposure to health-giving foods 334–6
　in Japan 307–15
　maintaining weight loss 329–32
　and motivational talk/reflective listening 321–5
　and older people 344–6
　outside motivation 319–26
　on personal level 316–19
　public health campaigns 326
　relapsers 330–1, 332–4
　removing barriers to change 326, 327–9
　and seamless change 326–7
　and sensory exploration/Sapere movement 337–46
　slimming magazines 316
　trying new foods 336–7
　see also diet
children's food
　bought by children 121–2
　and child pester-power 140
　and complex flavours 132
　connection with a person's future development 115–17

and convenience foods 133–4
difficulties in avoiding junk food
    141
digestible or indigestible 128–30
double standards concerning
    143–4
effect of class and money 122–3
exposure to SFS foods 146–7
family food 118–26
fruit phobia 124–6
as fun and different 134–9
gender differences 143–4
government involvement 112–13
haphazard nature of 120–1
high in salt, sugar and fat 137
innovations in 132–3
kid food 118, 134–45
as likeable or wholesome 117
mouth-cleansing foods 131
nursery food 118, 126–34
outgrowing 142, 144–5
purist attitude 141–2
rice pudding debates (1912–13)
    109–18
school food movement 112–18
as separate category 119, 127–32
and slum children 111–12, 114
taboos concerning 119–20
very hard or very soft 129–30
Children's and Household Tales
    (1812) 198
chilli 348–9
China 17, 160–1, 162, 163–6, 200,
    213, 253, 308
Chinese-Americans 180
chocolate 4, 21, 28, 47, 96, 146,
    155, 215, 221, 223–4
Cincinnati 111
Cinderella 197
Citizen Kane (film, 1941) 200
Clark University (Massachusetts)
    335
Codex Alimentarius 85

coffee 45, 46, 48, 75, 76–7, 78, 89,
    107, 108, 144, 165, 213
cognitive behavioural therapy
    (CBT) 276, 299
Coi: Stories and Recipes (Patterson)
    101
Coi restaurant (California) 101–2
Cooke, Dr Lucy 56–61, 62
Crabbe, Matthew 164
crisps 303–4
Crowley, Ralph 114–15, 116–17,
    121
Cuff, Marion E. 115–16
Culpeper, Nicholas 119

Daily Mail 112
David, Elizabeth 131, 144
Davies, Marvin 228–9
Davis, Dr Clara Marie 32–8, 39,
    42–3, 59, 64, 183, 341
Denmark 337
Devil May Cry 157
Dhaka 241
Diane (picky eater) 263–5, 300
Diego 277–8
diet
    and acquiring new tastes 115
    adolescent 22
    anxieties surrounding 1
    for babies 57–8
    balanced/healthy 13, 55, 61, 68,
        83, 163, 174
    barriers to change 327–34
    and better information 8–9
    calorie controlled 236–7
    in childhood 118–43
    and childhood memories 96–7
    in China 164–5
    crash diets 182
    crisis in 6, 7
    eating between meals 251
    extreme/limited 27, 54, 62, 264,
        272–83, 284–302

and flavours 45–6
global changes 145
hedonic shift 19
and hunger signals 255, 257, 258, 259
and idea of fullness 244
and images of desire 91
improving 105
insights into 352–6
and iron deficiency 212–13
Japanese 305–15
maternal 2, 80, 83, 105
misunderstanding of 5
narrowness of 19–20, 80
as nutritious 18
personal changes 316–19
pressures of 203–8
problems with 111
and seamless change 326–8
self-selection experiment 32–8, 40, 42–3
and sensory exploration 337–46
through motivational interviewing 321–6
tweaks to 335–6
Western 221, 307
yo-yo dieters 265–6
see also changing diets
disorders see anorexia; bulimia; eating disorders
Donald 34, 36
dopamine 15, 223
Drewnowski, Adam 332
Druckerman, Pamela 149
Dukes, Dr Clement 114
Duncker, Karl 64–9, 346
Duncker, Wolfgang 67
Dutch Hunger Winter (1944–5) 242
Dutch siblings 193–4
Dutton, Thomas 108, 126–7

EAT-26 test 220

Eating Disorder Not Otherwise Specified (EDNOS) 267
eating disorders
common safe foods 275
effect on family meals 293
experiences of 268–9
importance of family meals 301–2
inability to swallow 273–4
in later life 297–300
Maudsley approach (or family based treatment) 293–7
misconceptions concerning 265
picky eaters 263–5, 274–5, 303–4
Plate A-Plate B system 280–2, 346
preventing 283–4
problems of selective eating 271–3, 275–6
recovering from 266–7, 325
specific fear of food 273
taste exposure treatment 279–83
traditional treatments 276–7, 279
treatment of 269–70, 271
use of therapeutic paradox 278–9
variety of 265–6, 267–8
see also anorexia; bulimia
Economist 209
eggs 128, 213, 314–15
Enfagrow toddler milk 86
Engen, Trygg 88
Europe 210
Evening News 117

F100 239
fairy tales 197–200
Brothers and Sisters (Grimm fairy tale) 199
Hansel and Gretel (Grimm fairy tale) 198, 199
Snow White 197

family based treatment (FBT)
    293–7
family food 118–26
famine 162–3
fast food *see* junk food/fast food
Fat China (French & Crabbe) 164
fatty foods 5, 8–9, 105
feeding
    authoritarian style 178, 179–81
    authoritative 181–3
    and Baby Led Weaning 183–6
    and coercion/force-feeding
        169–76
    and division of responsibility 182
    as enjoyable activity 155–6
    and failure to thrive 153–4
    grandparents as overfeeders of
        children 160–6
    indulgent style 178–9, 181
    and leaving a clean plate 167–9,
        175
    link with child health 176–83
    offering treats 151–2, 153, 155,
        157–8, 159–60
    overfeeding/underfeeding 153–4,
        159, 160–4
    parental influence 153–6
    and self-regulation 183
    uninvolved style 177–8
    verbal/physical encouragement
        174–5
Feeding America 227
feeding disorders 268–70
    see also eating disorders
feta cheese 97
Fifth Sense conference (2014) 74
Finland 337, 338–41, 343
First World War 117
fish 33, 52, 61, 76, 121, 122, 128,
        133, 134, 189, 204, 207, 235,
        246, 247, 299, 305, 307,
        308–9, 310, 314, 336
Florida 229

food
    acquisition of eating habits
        18–19
    best diet confusion 9
    bliss point 5
    changing habits concerning 23–5
    cooking and growing 5–6
    eat more vegetables message
        9–10
    and eating disorders 6–7
    eating insights 352–6
    emotions associated with 154–5
    environment 16–19
    expressing parental love through
        166
    family influence 17–18, 20
    first order/second order
        preferences 20–1, 22
    'girly' or 'boyish' 210–12
    global food tastes 145–6
    hedonic shift 19
    low-fat products 8
    and obesity 5, 10–11, 13, 15, 17,
        19
    preferences for 3–4, 11, 21
    regulating sale of 24
    (re)learning how to eat 1–2, 11,
        14–15
    responding to hunger or fullness
        23
    scientific consensus on food
        habits 12
    self-perpetuating cycle 3–4
    structured mealtimes 23
    taboo things 238
    and taste for sweetness 12–14
    trying a variety of 23
Food Programme (BBC Radio 4) 78
Food Revolution (TV series) 334
Fourier, Charles 167–8
Fourier, Lubine 167–8
France 70, 92, 139, 149, 189, 203,
        210, 231, 337–8

French Institute of Taste 338
French, Paul 164
fruit 124–6, 134, 147, 181, 206, 218, 299, 327, 329

Gambia 163
'The Garden Party' (Mansfield) 149
gender
    and dietary guidelines/behaviour 218–21
    encouraging boys to eat 203–5
    and nutritional neglect of girls 195–7
    and response to food 209–10
    and sex-appropriate food 208–9, 210–15
    and sex-selective abortions 195
    and weight misperception 216–18
    and weight pressure 208
Gentleman's Relish 121
ghrelin 234, 235
Gigliotti, Luis 156–9
Glasgow 230
globus hystericus (lump in the throat) 273
glycaemic index (GI) 246
Gold, Rich 187
The Golden Cage (Bruch) 293–4
Goodnight Moon (Brown) 129
Gopnik, Adam 300
Grand Theft Auto 157
Grave of the Fireflies (film, 1988) 198–9
Greece 164
Greer, Germaine 173
Grimms' Fairy Tales 198
Guatemala 120
Guildhall (London) 111, 113, 115, 117

Hales, C. Nicholas 31, 163
Hall, Dr William 111–12

Hara hachi bu 244
Hardyment, Christina 126
Hare, Caspar 11
Hearst, George 200
Hearst, William Randolph 200–1
Henry, Diana 318
'Her Chee-to Heart' (McCorkle) 104
Hoban, Russell 51
Holt, Luther Emmet 128–9, 141, 171
honey 2, 21, 97, 237
hormonal biomarkers 233
Hostess Twinkies 104
Humble, Nicola 148–9
Humes, Michele 50
hunger
    effect on children 225–6, 229–30
    bottomless soup experiment 253
    and emotional eating 251–2
    food banks and breakfast clubs 227–8, 229–30
    glucostatic theory 233–4
    hormonal aspects 234–5
    and idea of fullness or satiation 243–50
    managing 226
    measuring and defining 230–7
    and mindfulness training 256–7
    Minnesota Starvation Experiment 236–7
    as negative concept 243
    numbers of people affected by 227, 229–30
    recovery from 229
    self-regulating 252–9
    and sensory-specific satiety 256
    size of portions/utensils 253–6
    subjective reports on 235–6
    symptoms of long-term hunger 228–9
    time between meals 250–1

topping up with snacks 226–7,
    231–2, 251
treating 238–43

ice cream 4, 129, 144, 148, 155
India 195, 240, 242, 328
International Centre for Diarrhoeal
    Disease Research (Bangladesh)
    243
Ishige, Naomichi 314
Itard, Dr 337–8
Itoh, Makiko 188

Jane (anorexic woman) 297–8
Japan
    changes in diet 311–15, 343
    and concept of 'delicious' 310
    eating of ramen in 309–10
    famine in 198–9
    female drinking and eating
        208–9, 210, 211
    healthy attitude toward eating
        305–7
    obesity in 306
    poor cuisine in 308–9
Johns Hopkins University 256
Johnson, Susan L. 257–8
junk food/fast food 12, 13, 20, 21,
    29, 40, 49, 50, 82–3, 103–6,
    181, 301, 306, 316, 341
Just Bento (Itoh) 188
Jyväskylä 340–1

Kalymnos 98
Katz, David L. 8–9
Kawash, Samira 136
Kayman, Susan 331, 332
Keyworth Primary 228, 229
KFC 208
kid food 118, 134–45, 147
Korea 308, 310
Köster, E. P. 23, 71, 95–6, 342
Kraft Foods 138

Kunming 253
Kushner, Barak 308–10, 311
Kuwait 17, 219–21

Lambeth (London) 122–3
Laos 2
Leach, Penelope 133–4
Leeds 111
Leigh, Rowley 117
leptin 234–5
'Let's Move' programme 141
Levi, Primo 93
Lewis, Morris 94
Libya 219
likes/dislikes
    acquisition of preferences 51–2
    bizarre 27–8
    and children with learning
        difficulties/other disabilities
        61–4
    and disgust 43–5
    diverse 28–9
    and flavour window in babies
        57–9
    genetic/epigenetic argument
        30–2, 41–3, 44, 49, 50, 55–6
    mere exposure thesis 50–1, 52
    and neophilia 53–4
    and neophobia 52–6
    PROP tasters 45–50
    self-selection experiment 32–8,
        40, 42–3
    and social suggestion 64–9
    Tiny Tastes approach 59–61
    twin studies 41
    wanting/liking difference 29–30
    wisdom of the body view 38–40
Lily 336–7, 342
Little Chef 236
Little House on the Prairie (Wilder)
    159–60
Lock, James 295–6
London Hospital 131

Lowinsky, Ruth 110
Lunchables 138
lunchbox 187–8
Lustig, Robert 13–14
Lyytikäinen, Arja 340, 343–4

macaroni cheese 99, 105, 112, 204, 252
McConnell, Carmel 228
McCorkle, Jill 104
McDonald's 141
McMillan, Margaret 114–15
Magic Breakfast 228, 229
Malawi 194, 239
malnutrition 10, 114, 162, 194, 227, 229, 238–43
Manary, Mark 239–40
Mansfield, Katherine 149
Mao Zedong 162
marshmallows 102–3, 136
Massachusetts General Hospital 288
Maudsley Hospital (south London) 293
Mayans 120
Mayer, Jean 233
Mead Johnson 86
Mead, Margaret 143
meat 8, 16, 20, 30, 32, 52, 54, 61, 91, 98, 106, 112, 119–20, 121, 122, 127, 128, 129, 141, 143, 159, 161, 174, 188, 193, 203, 204, 209, 210, 211–12, 213, 214, 219, 247, 275, 299, 311, 312, 314
Medici children 119
medium tasters 45
memory
    carrot juice study 83
    childhood 79–80, 96–9, 104–6
    conscious/unconscious 78–9
    convenience/packaged foods 103–6

culturally determined differences 100–1
    and desensitization 103
    and development of taste buds 81–2
    and flavour-yearning 89–95
    in foetuses and babies 81–7
    and gastronomic desires 95–6
    importance of 78–9
    and olfactory discernment 87–9
    problems for chefs 99–100, 101–3
Men Love Pies, Girls Like Hummus (Rimmer) 210
Mendelson, Charlotte 98
Mennella, Julie 83, 84, 100
mere exposure 50–1, 52, 279
Middle East 210, 220
Military Diet Research Committee (Japan) 312
milk 2, 27, 32, 33, 34, 37, 62, 85, 107–8, 113, 120, 123, 131, 139, 142, 143, 144, 152, 225, 228, 260, 268, 288, 296, 303, 311, 312, 313, 347
    breast milk 2, 39, 57–8, 83–5, 119, 171–2, 185
    dried 239, 243
    formula milk 2, 58, 84–7, 154
milk pudding *see* rice (milk) pudding
milkshakes 165, 248, 297, 301
Millard, Abbi 73–5, 77
Miller, William 321–2
Minnesota Starvation Experiment (1944–5) 236
mint tea 92–3
Mobley, Amy 179
Momofuku restaurants (New York & Toronto) 144
Monell Chemical Senses Center (Philadelphia) 83, 100
Moss, Michael 5

motivational interviewing 321–5
Motivational Interviewing (Miller
    & Rollnick) 321–2
MSG 86, 101, 249
Musaiger, Abdulrahman O. 220–1

Nathan's (Coney Island) 106
National Center for Women's and
    Children's Health (Beijing) 165
National Health Service (NHS) 216
National Research Council's
    Committee on Food Habits
    143
neophilia 53–4
neophobia 52–6
Nesbit, E. 110
Nestle, Marion 86, 254
Netherlands 337, 338
Never the Same (film, 2012) 94
New York City 121, 122, 180, 307
New York Times 106
New York University 254
New Yorker 300
Newmarket House (Norwich) 298
Nicholls, W. A. 114
Nigeria 169
No Kid Hungry 225–6
non-tasters 45, 46, 47, 48, 49
North America 210
Norway 339
nursery food 118, 126–34
Nutella 239

Obama, Michelle 9, 141
obesity
    and being given wrong advice 8
    in children 17, 18, 41, 48, 108,
        215–19, 243, 340
    in China 164–5
    effect of dopamine 15
    epidemic 13
    and feeling full 245
    and genetic inheritance 194

in Japan 306
judgmental impatience
    concerning 10–11
in Kuwait 219–21
and low income 19
and motivational interviewing
    324
in rich countries 5
rising rates of 8
Oliver, Jamie 141, 334
The Omnivorous Mind (Allen) 304
One and Only (Sandler) 201
Ottolenghi, Yotam 44, 48
overeaters 22

Pac-Man 156, 181
Pakistan 328
Paltrow, Gwyneth 246–7
Pande, Rohini 196–7
Patterson, Daniel 101–3
Pearson, Dympna 319–26
Pellegrini, Angelo 132
Penn State Hershey Children's
    Hospital Feeding Program 140,
    272, 279, 281
Penn State Hershey Medical Center
    (Pennsylvania) 61–2, 63
Pennsylvania 175, 252–3
The Physiology of Taste (Brillat-
    Savarin) 204
Pizza Hut 220
Plate A-Plate B system 280–2, 346
Plumpy'Nut 239–43
Pollan, Michael 10, 165
Pooley, Siân 124
Popkin, Barry 145, 250
Portsmouth 114
post-ingestive conditioning 15
Praader-Willi syndrome 235
preloads 247–8
Prescott, John 53
prisoners of war (POWs) 93–5
Pritchard, Dr Eric 130, 134

Project EAT (Minnesota) 206
PROP tasting 45–50
Proust, Marcel 80, 103, 104
Provision of Meals Act (1906) 112
Puisais, Jacques 338
Purdue University 250

Ragazzi chain 135
Rai, Baldeesh 163
The Railway Children (Nesbit) 191
Rainey, George 114
ramen 309–10
Rapley, Gill 183–5
Ready to Use Therapeutic Food
    (RUTF) 239–43
Reeves, Maud Pember 122–3
Renaissance 124
retsina effect 28–9
rice 2, 42, 57, 93–4, 98, 101, 242,
    243, 307, 313, 314
rice (milk) pudding 109–18, 130,
    131, 133, 243
Rimmer, Simon 210
Ringwood, Susan 290–2
Robinson, Jancis 46
Roden, Claudia 219
Rollnick, Stephen 321–2
Rolls, Barbara 256
Rozin, Elizabeth 315
Rozin, Paul 28, 43–4, 347–8
Rugby School 114

salad 19, 33, 34, 68, 69, 71, 74,
    102, 119, 128, 144, 214, 317
San Lorenzo (Florence) 119
Sandler, Laura 201
Sapere movement 337–46
Satter, Ellyn 182
Scheindlin, Benjamin 38
Schiller, Deborah 347–8
school lunches 187–8
Second World War 143, 313
sensory education 337–46

SFS (sugar/salt/fat) 137, 145–6,
    341
Shephard, Sue 94
Shepherd, Gordon M. 90
Shozo, Marumoto 312
siblings
    and fairy tales 197–200
    favouritism 194–5
    gender differences 195–7, 202–8
    genetic inheritance 194
    influence of companions 192–3
    older ones copy younger ones
        193–4
    and the only child 200–2
    parental bias 205–8
    potential of 221–2
    power over eating habits 192
    rivalry over food 189–92
    similarity in eating habits 193
Slater, Nigel 208
Spain 224
Somers Town nursery school
    (London NW1) 65
Song of Solomon 166
soup 22, 111, 114, 127, 166,
    175–6, 204, 207, 211, 245,
    246, 248–50, 253, 298, 305,
    306, 307, 311, 314, 315, 335,
    353
South Asians 328
Southern Methodist University
    (nr Dallas) 174
Spargo, John 121
Spieler, Marlena 75–6, 78, 93
Spock, Dr Benjamin 38, 133, 336
Steiner, Jacob 81
Stewart, Dr Laura 215–16
The Story of the Treasure Seekers
    (Nesbit) 110
Studio Ghibli 198
supertasters 45–50
Sutton, David 98
Swarthmore College 69

Sweden 161, 213, 337, 338, 339
sweet foods 4, 5, 10, 13, 39–40,
    52, 82, 86–7, 94–5, 105, 114,
    133, 136, 139, 144, 146–7,
    148–9, 151–2, 155, 156–7,
    159–60, 165, 172, 173, 181,
    221, 246, 248, 335
Syria 219

Tanzania 2, 120
Tayside 215–16
Thailand 218
Thompson, Jan 94
Thoreau, Henry David 126
The Times 168–9
Times of India 195
Tiny Tastes approach 59–61
Toast (Slater) 208
Tokyo 305
Tuorila, Hely 342
twin studies 41, 194, 257, 290
Tyler 281–2

Ulander, Kerstin 346
University College London 56
University of Kentucky 329
University of New Mexico 322
Urbick, Bryan 210, 223

vanillin 85–8
VE (anorexic girl) 288–90
vegans 213
vegetables 2, 4, 9–10, 11, 12,
    22, 39, 44–50, 52, 54, 56–7,
    58–61, 70, 128, 129–31, 134,
    167, 174, 181, 206, 214, 218,
    246, 247, 304, 317–18, 327,
    329

broccoli 4, 9, 47, 49, 52, 58, 74,
    91, 132, 134, 138, 172, 195,
    246, 333
Brussels sprouts 12, 44, 45, 46,
    49–50, 145
vegetarians 213–14
Veronese, Paolo 204
Victor 337–8
video games 156–8
Virginia 120
Vollmer, Rachel 179

Wallace, Dr 131
Wansink, Brian 254–5
Wardle, Jane 56
Washington, Booker T. 120–1
Weber, Eugen 198
weight problems 207–8, 213–18,
    318
Whittome, Susi 228
Wilder, Laura Ingalls 159–60
Williams, Dr Keith 63, 140, 272,
    274–5, 278–83, 346
Williams, Zoe 141
Wise, Roy A. 79
World Health Organization
    (WHO) 57, 85, 183, 212
Wright, Charlotte 185

Xiamen 160–1

Yale University 45
yoghurt 21, 39, 74, 108, 133, 138,
    139, 149, 173, 317
Your Baby and Child (Leach)
    133–4

Zajonc, Robert 50–1, 279